301.29   77151
F77t

DATE DUE		
Nov 6 '72		
Nov 20 '72		

TIME AND SOCIAL STRUCTURE
AND OTHER ESSAYS

LONDON SCHOOL OF ECONOMICS
MONOGRAPHS ON SOCIAL ANTHROPOLOGY

Managing Editor: Anthony Forge

The Monographs on Social Anthropology were established in 1940 and aim to publish results of modern anthropological research of primary interest to specialists.

The continuation of the series was made possible by a grant in aid from the Wenner-Gren Foundation for Anthropological Research, and more recently by a further grant from the Governors of the London School of Economics and Political Science. Income from sales is returned to a revolving fund to assist further publications.

The Monographs are under the direction of an Editorial Board associated with the department of Anthropology of the London School of Economics and Political Science.

LONDON SCHOOL OF ECONOMICS
MONOGRAPHS ON SOCIAL ANTHROPOLOGY
No. 40

TIME AND
SOCIAL STRUCTURE
AND OTHER ESSAYS

BY

MEYER FORTES

UNIVERSITY OF LONDON
THE ATHLONE PRESS
NEW YORK: HUMANITIES PRESS INC.
1970

Published by
THE ATHLONE PRESS
UNIVERSITY OF LONDON
at 2 Gower Street London WCI

Distributed by Tiptree Book Services Ltd
Tiptree, Essex

Australia and New Zealand
Melbourne University Press

Canada and U.S.A.
Humanities Press Inc.

© *Meyer Fortes*, 1970

U.K. 485 19540 2

U.S.A. 391 00112 4

Printed in Great Britain by
WESTERN PRINTING SERVICES LTD
BRISTOL

PREFACE

The papers reprinted in this volume have been selected with two considerations in mind: they record ethnographical observations from my field work among the Tallensi and in Ashanti that are not easily accessible elsewhere but continue to be useful for comparative studies and as background to current research in Ghana; and they represent applications of methods of analysis and schemes of interpretation that were emerging in British structural anthropology at the time of their publication. What they have in common is the endeavour to bring out the mechanisms and processes of social structure that lie behind and are reflected in the customary arrangements and forms of behaviour described.

The vast expansion of field research, and the advances in theory since the middle forties, have so changed the state of our knowledge in social anthropology that it might be useful to give some indications of the circumstances in which these papers were written. The paper on Tallensi ritual festivals (Chapter 6, first published in 1936) arose out of my first period of field work in 1934-5. Up to that time no intensive field research had been carried out in any of the tribal groups of what were then the Northern Territories of the Gold Coast. Rattray, in his ethnographical survey of the area (1932, **ii**, p. 358) mentions having heard of these festivals, but, as he told me himself, he never attended them. I was fortunate enough, through the good offices of the chiefs and tendaanas responsible for the ceremonies, to take part in both the public and many of the private celebrations and rites during my first period of field work. They made a great impression on me, in particular, by their patent importance for maintaining a framework of social and political integration in what seemed at first sight to be a politically amorphous cluster of independent lineages. My paper was intended to show how these ostensibly 'magico-religious' (as Marett called them) celebrations served, in fact, to dramatize the political interdependence of the two major groups of clans which, in other contexts, emphasized and paraded their genealogical, local and religious distinctness and opposition. The pattern of obligatory ritual collaboration (which could, incidentally, only have been discovered by observation in the field) was obviously the key; and analysis of it revealed the elaborate scheme of structural polarities embodied in the ceremonies. I became aware thus, of the part played by complementary opposition in Tale social structure and politico-ritual ideology. However, the full implications of this only became clear to me later, when I set about the analysis of the lineage system (1945).

I might add that when I revisited Taleland in 1963, I found the Harvest Festivals in full swing and learnt that the Golib Festival continues to be celebrated in traditional style. But nowadays, I was told, a great multitude of spectators coming from as far afield as Kumasi and Accra, attends the dances. The opening up of the North, as a result of modern geographical and occupational mobility and the extension of the road system, has resulted in a great deal of publicity being given to these ceremonies.

The study of the indigenous educational practices and customs of the Tallensi (Chapter 8, first published in 1938) stems from my second period of field work, in 1936–7. Though my interest in this topic reflects, in part, my earlier training in psychology, the investigation was prompted, primarily, by my endeavours to understand the relationships of parents and children in the context of Tale culture and family structure. Malinowski's ideas about the development of family relationships and the transmission of the sentiments and attitudes involved in them, were an added stimulus. There were, at that time, no schools in Taleland. The handful of boys who were receiving a school education attended schools outside the tribal area, as boarders. Today, of course, the majority of the children of school age attend schools up to the secondary level in their own localities and a substantial number in the older age groups are away at secondary boarding schools or teacher training colleges or the universities. The patterns of home-based training I observed in the thirties are still met with in the treatment of pre-school children and in the inculcation of the sentiments and values of kinship and family relationships. An index of the attachment Tallensi have to their traditional kinship values is the practice, followed by many migrants residing in the towns of Southern

Ghana, of sending their children home to live with a brother and attend school in Taleland, avowedly with the aim of ensuring that they learn something of the traditional way of life. But modern schooling is oriented to an economic and political environment and to ways of thought that are radically different from those of the past. As literacy, already common among the younger Tallensi, and school-bred habits of life and of thought spread, many of the traditional practices described in this paper will disappear and the observations recorded in it will not be repeatable.

The paper on 'The significance of descent in Tale social structure' (Chapter 2, first published in 1943) is, in effect, a synopsis of the analysis of the lineage system presented at length in *The Dynamics of Clanship*. In attempting thus to set up a paradigm of the general principles on which the lineage system is based, I wanted, in particular, to draw attention to the dynamic equilibrium inherent in it and to show how it arises, as a process in time, out of the complementary interplay of patrilateral and matrilateral descent relationships rooted, ultimately, in the structure of the family.

I return to the problem of the time factor in kinship and family organization in 'Time and Social Structure' (Chapter 1, first published in 1949). The data derive from my field work in Ashanti, in 1945-6, in collaboration with Mr (now Professor) R. W. Steel and Miss Peter Ady (*cf.* Fortes, Steel and Ady 1948). My intention was to supplement the meagre and broadly generalized description of the Ashanti family given by Rattray (1929, pp. 1–32; *cf.* Fortes 1950) with an analysis of its contemporary structure. But it was evident that the conventional descriptive terminology for types of residence and domestic organization would not make sense of the data. To resolve the apparent inconsistencies it was necessary first to isolate the variables of sex, age, familial status and kinship relationship that enter into the structure of the family, and then to recognize that the descriptive diversity of the observed domestic arrangements is correlated to stages in a developmental cycle. I had previously used this frame of analysis to account for variations in the synchronic constitution of Tallensi domestic families (1949a, pp. 63–77) but the Ashanti situation, as I found it in 1945, was much more complex. The interaction of the more differentiated external social and economic circumstances and the internal, still mainly traditional norms of kinship and domestic organization raised problems of analysis not represented in the Tallensi model. This required consideration of some basic conceptual and methodological matters. The case for the indispensability of the numerical data adduced needed stating and the significance I attached to temporal extension and sequence for the understanding of social structure had to be explained. This led me to contrast the 'qualitative' aspects and the actually or ideally 'quantitative' features of social organization and to suggest that this corresponds to the dichotomy of 'culture' and 'social structure' as we ordinarily use these terms.

The approach to family and kinship systems tried out in this paper has proved useful in a number of subsequent studies (cf. e.g. R. T. Smith 1956; the papers collected together in Goody, J. (ed.) 1958; and those in Gray and Gulliver (ed.) 1964). The application of numerical methods in this paper now looks amateurish in comparison with current quantitative and statistical research in social anthropology (cf. e.g. Buchler and Selby 1968, with particular reference to kinship). Of most interest, perhaps, is the attention that has been paid to the conceptual problem of the relationship between normative rules and the choices actually made by individual (cf. Leach 1960, in Murdock, ed., 1961, ch. 1).

The papers on 'The Structure of Unilineal Descent Groups',[1] on 'Analysis and Description in Social Anthropology' and on 'Descent, Filiation and Affinity' (Chapters 3, 5 and 4) supplement one another. They outline an analytical approach to the study of kinship and social organization which I develop at length in a forthcoming book (1969). The analysis of descent group structure and its familial and politico-jural concomitants presented in these papers has attracted frequent discussion, so much so that I have thought it worth while to add an appendix reviewing the most important of recent contributions to the subject.

The review article on Radcliffe-Brown (Chapter 9) needs no further comment. And all that needs to be said of Chapter 7, which reprints my Henry Myers lecture, is that it is one of a

[1] This paper has once before been reprinted in Ottenberg, Simon and Phoebe, (editors) *Cultures and Societies of Africa*, Random House, New York, 1960, pp. 163–89.

series of studies in which I have drawn on the findings of psychoanalysis to elucidate some aspects of Tallensi ancestor worship
(1959, 1964).

I am indebted to Professor Isaac Schapera and to Mr Anthony
Forge for advice and editorial assistance in bringing out this
collection of papers. I am obliged also to the editors and publishers
of the following journals and books for permission to reprint
these papers:

The American Anthropologist; *The Advancement of Science*; The
Clarendon Press, Oxford; *The British Journal of Sociology*; *The
Journal of the Royal Anthropological Institute*; *Man*.

Postscript: While correcting the page proofs of this book, my
attention was drawn, by Roxanne Gudeman, to the reference
made to the paper on 'Social and Psychological Aspects of Education in Taleland' by Dr Jerome S. Bruner, the distinguished
authority on the psychology of thinking and cognition, in a
recent work (Bruner, Jerome S., Oliver, Rose, R., and Greenfield,
Patricia M., 1966, *Studies in Cognitive Growth*, New York, pp.
60–2). In following up and extending the methods and theories
of Piaget, in their studies of cognitive development among children, Bruner and his colleagues find it valuable to take into
account cross-cultural observations and experiments; and it is in
this connection that my observations are cited by them. Dr
Bruner's generous assessment of the usefulness of my data is
reassuring. An earlier reference to my observations is to be found
in a basic study of the social and cultural factors in learning, first
published during the war (Miller, Neal E. and Dollard, John,
1941, English edition 1945, London, *Social Learning and Imitation*,
p. 263). The authors describe my account of imitative behaviour
as 'very sensible' and in line with their own reinforcement theory.

CONTENTS

PLATES

(between pages 252–3)

I

Time and Social Structure:
An Ashanti Case Study[1]

In his most recent discussion of the concept of social structure Radcliffe-Brown (1940a) distinguishes between 'structure as an actually existing concrete reality' (that is, 'the set of actually existing relations, at a given moment of time, which link together certain human beings'), and 'structural form' (that is, 'the general or normal form' of a relationship 'abstracted from the variations of particular instances, though taking account of these variations'). The distinction is associated with the 'continuity of social structure through time' particularly in a 'relatively stable community'. The structural form may change little, though the actual social structure is constantly being renewed and changes with the birth and death of members and their changing relations with one another. Developing his thesis, Radcliffe-Brown raises the elementary but fundamental question: How is the 'norm' established?

The distinction seems to be of doubtful validity. It errs, I believe, by attempting to synthesize a number of separate theoretical issues in one formula. To begin with, the time factor in social structure is by no means uniform in its incidence. Briefly, the following functions of time can be distinguished:

(a) As mere *duration* time is an extrinsic factor having no critical influence on the structure of social events or organization. Conversations, many ceremonies, court cases, the activities of a fishing group based on a canoe (cf. Firth 1946, p. 114), all occur over a stretch of time but are not intrinsically determined by this fact. Distribution and location in time are aspects of

---

[1] Reprinted from *Social Structure: studies presented to A. R. Radcliffe-Brown* (M. Fortes, ed.), 1949.

duration, as we see in seasonal activities, annual ceremonies, or institutions like the kula (Malinowski 1932a). Simple sequence is another aspect of duration, as most history books show.

(b) As *continuity* (or its opposite, discontinuity), time is an intrinsic and critical characteristic of some social events or organizations. In this case it is significant as an index of forces and conditions that remain more or less constant over a stretch of time; or else of those that give way precipitately to new forces and conditions. Thus all corporate groups, by definition, must have continuity. A lineage in a stable and homogeneous society exemplifies this (*cf.* Fortes 1945, ch. 3). It is kept in being by, and is at a given time an expression of the forces that determined its structure in the past and will do so in the future. Discontinuity is exemplified in areas where military invasion, conquest, and alien settlement on a big scale have occurred, or where a population of very diverse origins has been brought together forcibly in an alien land, as was the case with African slaves in the New World (*cf.* Simey 1946).

(c) Finally, time may stand for what might be called *genetic* or growth processes, as opposed to mere historical sequence. Time is then correlated with change within a frame of continuity. Growth in this sense is more marked in social systems that are not in a more or less stable equilibrium than in those that are, though it occurs in the latter too. For growth is the product of two kinds of forces symbolized by the passage of time, those of continuity (conservative forces) and those of non-reversible modification. It may appear as simple increase or decrease, as in population changes, or in more complex qualitative differentiations in institutions and social relations. These may be of a negative kind due to contradiction, involution, or loss of parts or functions; or they may be positive, showing expansion, accumulation, or development of parts or functions. The British House of Commons is a familiar instance of growth in social institutions and organization (*cf.* Jennings 1939, 1948).

It is pertinent to add that what has been said of time applies equally to the spatial factor which is universally present in social structure (*cf.* Evans-Pritchard 1940a, ch. 3). Thus we find mere location in space, which is extrinsic to the event or organization; ordered arrangement in which spatial relations directly shape structure; and controlled movement representing change in structure.

SOCIAL STRUCTURE

I take 'structure' to refer to a distinguishable whole (an institution, a social group, a situation, a process, etc.) which is susceptible of analysis, in the light of appropriate concepts and by suitable techniques, into parts that have an ordered arrangement in space and time. Of course, what is a 'whole' in one context may be a 'part' in another and may be resolvable into an arrangement of less complex 'parts' as theory develops. What is really important, however, is not merely the determination of the 'parts' and their interrelations but the elucidation of the principles which govern structural arrangement and of the forces for which these stand.

In my view the distinction between 'actual structure' and 'structural form' is invalid, because structure is not immediately visible in the 'concrete reality'. It is discovered by comparison, induction, and analysis based on a sample of actual social happenings in which the institution, organization, usage, etc., with which we are concerned appears in a variety of contexts. When we describe structure we are already dealing with general principles far removed from the complicated skein of behaviour, feelings, beliefs, etc., that constitute the tissue of actual social life. We are, as it were, in the realm of grammar and syntax, not of the spoken word. We discern structure in the 'concrete reality' of social events only by virtue of having first established structure by abstraction from 'concrete reality'.

The real problem is that in social structure we are always faced with parts and relations of diverse nature and variability. There may be parts and relations which recur in all situations in which the organization or institution we are studying emerges, and others which seem to occur only by chance. The former may be constant in some respect but variable in others, corresponding in some ways to the statistician's independent variable; the latter may be a 'normal' usage or institution in an 'abnormal' context. The chief difficulties arise because of the great lack of uniformity in the nature of what we discern as a constant element. It may be what persists as opposed to what varies over time, so that 'constant' means the frame of continuity and 'variable' the process of growth or change. Thus among the constant features in the structure of the Tale lineage is its division into precisely defined segments, and among the variable features is the number

TIME AND SOCIAL STRUCTURE:

of such segments (Fortes 1945, ch. 12). Or again, 'constant' may refer to what is considered essential or intrinsic, 'variable' to what is considered incidental. In the African institution of the bride-price, the passage of goods or valuables from the bridegroom's side to the bride's side is a constant, that is, essential, feature; but the valuables used and the amount passed vary widely even in one society.

If we consider the operations required to find valid answers to our questions, it appears that social behaviour in its collective aspects yields two kinds of data. On the one hand, we have data which have meaning primarily in terms of magnitude. The amount of bride-price actually paid; the range within which classificatory kinship is recognized; the generation depth of a lineage; the extent of observance of a legal or moral rule; and many similar data of social relations require investigation in quantitative terms,[1] though we still lack the techniques for dealing with all of them in this way. On the other hand, the obligation to pay a bride-price in a fixed customary way; the social recognition of matrilineal descent; the belief in witchcraft; and so forth, are data which can only be dealt with by direct apprehension and qualitative description. Both the parts and the relations between them which we discriminate in any study of social structure have these two sides. No doubt as social anthropology develops, more and more of what can at present only be understood in qualitative terms will be broken down into parts and relations susceptible of quantitative analysis.

The qualitative aspect of social facts is what is commonly called culture. The concept 'structure' is, I think most appropriately applied to those features of social events and organizations which are actually or ideally susceptible of quantitative description and analysis.[2] The constant elements most usually recognized in any social event by ethnographers are its cultural components; it structural aspect, being variable, is often overlooked. It should be emphasized that I am not suggesting a division of the facts of social life into two classes; I am referring to the data of observation. 'Culture' and 'structure' denote complementary ways of analysing the same facts. In the present stage of social anthropology all analysis of structure is necessarily hybrid, involving descriptions of culture as well as presentation of structure. The

[1] I include in this geometrical or quasi-geometrical methods and concepts.
[2] This is much the same point as was made by Bateson (1936, ch. 3).

factors of space and time are of fundamental importance in this connection. Structure is given to social happenings or organizations by space and time relations, if for no other reason; and these relations have magnitude.

METHODS

We can now consider the problem of the methods by which the 'norm' is established. The concept of a 'norm' needs more precise definition than Radcliffe-Brown gives to it. It may stand for what I have called the constant elements; or it may be equivalent to the statistician's mean or mode; or to the lawyer's precedent, or the moralist's ideal pattern. Whatever sense is given to the term, practice as well as logic proves that a 'norm' can seldom be satisfactorily established from a single instance.[1] Even events that are unique for most ethnographers, such as a ceremony in a cycle that takes a generation,[2] can only be satisfactorily studied by comparing their component activities, beliefs, arrangements, etc., with similar social elements occurring in other contexts in the life of the people. There is plenty of evidence to justify the assumption that the modes of behaviour and principles of organization current in a society are limited in number and variability. But though it is the common practice among ethnographers to arrive at the 'norms' by comparison of and induction from repeated instances, this is rarely if ever done in relation to explicit criteria of verification and validation.

Satisfactory as these methods may be in stable and homogeneous societies they lead to serious errors in societies of diversified structure. Institutions which in a stable and homogeneous society adhere together as closely interdependent variables tend to be mutually independent, so that a wider and less predictable range of combinations is possible for them in different situations. Where in a stable and homogeneous society almost all social relations may be governed by one or two general principles,

---

[1] In theory, of course, a single case can be an adequate sample of a perfectly homogeneous population, and a great deal of ethnographic information is perfectly reliable even though it may be based on a single case or on the statements of one informant. The only check needed is on the truthfulness of the informant. Identification of the language family to which the speech of a given community belongs can be based on the speech of one truthful informant.

[2] e.g. the *Maki* rites of the Vao of Malekula. (*Cf.* Layard 1942, chs. 11–12.)

such as kinship, in a highly diversified society there will be a large number of such principles, none of them quite general. In these societies 'norms' cannot be discovered by inspection or haphazard comparison. More systematic methods are necessary, and that means the application of statistical concepts. This is taken for granted by students of European and American societies; and the value of statistical methods, of however elementary a nature, for the study of so-called simpler societies, has also been demonstrated by anthropologists (cf. Firth 1946; Schapera 1943), but it is not generally accepted that they are essential for the study of social structure in all societies and that in fact they are nothing more than a refinement of the crude methods of comparison and induction commonly used. Statistical techniques are not essential and are perhaps even inappropriate in the study of social life from the cultural aspect.

The application of statistical concepts will show that the concept of 'structure' is most appropriately used for the kind of abstract or generalized description which Radcliffe-Brown calls 'structural form'. The concept 'form' is widely and somewhat indiscriminately used by ethnographers and social anthropologists. We speak of the form of the family or of marriage as being monogamous or polygamous.[1] Benedict (1935, ch. 1, *passim*) uses 'cultural forms' as the equivalent of 'patterns of culture'. The Wilsons (1945, ch. 3) appear to regard *structure* and *form* as synonymous, structural form being contrasted with the cultural content of social relations. I myself (1945, ch. 13) have described the Tallensi as having a society of segmentary form by contrast with societies having a pyramidal form of political structure. In general these uses of the term 'form' suggest that it might be restricted to refer to those characteristics of an arrangement of parts that distinguish it in its totality. Thus a subsistence economy based on fishing might be similar in form to one based on agriculture though there is a great difference between them in respect to their component parts.[2]

THE PROBLEM OF ASHANTI DOMESTIC ORGANIZATION

The importance of the methodological problems raised by Radcliffe-Brown can be illustrated by considering domestic

---

[1] *Cf.* MacIver (1937, p. 214): 'The Family has no one original form in the sense of a specific primal type. . . .'   [2] As Firth argues, *op. cit.*, p. 22.

organization in modern Ashanti.[1] I cannot here deal in detail with this subject, but I select one or two main problems for discussion.

In modern Ashanti, as elsewhere in Africa, social ties based on the recognition of genealogical connections have a dual function in ordering social relations. On the one hand, matrilineal descent exclusively determines membership of the lineage. It is the localized, corporate group forming the basis of all political, jural, and ceremonial institutions. Membership of the matrilineage is *ipso facto* membership of a widely dispersed, exogamous clan.[2] Lineage and clan depend for their existence on continuity in time.

On the other hand, however, ties of kinship, marriage, and affinity regulate the structure of domestic and family groups, which have no permanent existence in time. Each domestic group comes into being, grows and expands, and finally dissolves. But the institutions it embodies, and the mode of organization it exhibits, are essential features of the social structure. Domestic organization has two aspects. Its form derives from a paradigm or cultural 'norm' sanctioned by law, religion, and moral values. Its structure is governed by internal changes as well as by changing relations, from year to year, with society at large.

Ashanti to-day is not a stable and homogeneous society. Occupational differentiation; stratification by income, education, and rank; geographical and social mobility; as well as disparate values in religious belief, morality, law, and personal ideals, have produced a diversified and in parts unstable social system. One sign of this is that there *appears* to be no fixed norm, of domestic grouping. The influence of processes of growth correlated with age, sex, social maturation, marriage, economic achievement, and so forth is marked in the structure of the domestic group. Thus the usual ethnographic method of describing domestic organization is not applicable.[3] Such 'blanket'

[1] The data used in this paper are part of the results of the Ashanti Social Survey, 1945–6, a preliminary account of which is given in Fortes, Steel, and Ady (1948). Some of the statistical data cited are derived from census records collected in the field by Miss P. Ady, to whom I am also indebted for valuable suggestions for their analysis.

[2] The *loci classici* for Ashanti social organization are of course the splendid works of R. S. Rattray: 1923, 1927, 1929. I shall, however, only cite Rattray if there is disagreement between my own field observations and his statements.

[3] It is of interest that Rattray, who gives very detailed, and as far as I have been able to ascertain, accurate information on kinship relations (1923 and 1929), says

terms as patrilocal and matrilocal are quite useless. As in our own society,[1] more rigorous methods of a statistical kind are necessary.

THE DATA USED

I shall use data from two rural areas in Ashanti, about thirty miles apart as the crow flies but not linked by modern roads. The social and cultural changes of the past thirty to forty years have led to variations, often of considerable extent, between one part of Ashanti and another. The two areas selected appear to include most of the variations observable in domestic organization outside the large towns and industrial centres.

The areas selected are the townships of Asokore and Agogo. Both places are the headquarters of subordinate chiefdoms of the Ashanti Confederacy. Such a chiefdom is under the authority of its own chief and his councillors, all locally elected. It consists of the capital township and a surrounding stretch of territory in which there are usually villages, hamlets, and scattered farms occupied by subjects of the chiefdom, very many of whom often have homes in the capital township as well.

The two chiefdoms are comparable in many respects. The economy of both is based on cocoa production, which reached them about forty years ago; but Asokore is in an area of declining prosperity, whereas Agogo has maintained a higher standard of living. Missions and schools, modern commerce, British rule, and all the other agencies of Western influence have had about equal effect in both places. Between half and two-thirds of the people of both places claim to be Christians. A similar range of occupations and professions, varying from school-teachers and clerks at one extreme to immigrant sanitary labourers at the other, is found in both townships. One significant difference between them is in their geographical situations. Asokore lies in a thickly populated area twenty-five miles from Kumasi, the capital and largest modern city (pop. 70,000) of Ashanti, on an excellent motor road. Agogo is at the end of a second-class road eighteen miles from the mining centre of Konongo. Urban contacts and in-

almost nothing about domestic organization. In 1929, p. 9, and in 1927, p. 326, he implies that marriage and the family are patrilocal.

[1] And in other societies undergoing rapid social differentiation. Schapera and, following him, Ashton have shown that the conditions described above also apply among the Southern Bantu to-day. See Schapera (1935) and Ashton (1946).

fluences are therefore much stronger at Asokore than at Agogo.

More important are the differences between the two places in their political geography and recent history. Asokore township has a resident population of about 900. This is about a quarter of the population of the chiefdom, most of which is scattered among the outlying villages and hamlets. The population of Agogo township is just over 4,000, and this includes about two-thirds of the population of the chiefdom. As this suggests, Agogo is a more closely knit community than Asokore. Comparing the two places, the observer cannot help noticing that there is, at Agogo, a sense of unity and of corporate identity associated with pride in the achievements and antiquity of the chiefdom, which is missing at Asokore. Yet both townships are known to have been in existence for a long time, certainly not less than a century. Indeed genealogies are cited to show that the founding ancestors of the lineages occupying the townships at the present time came there some ten to twelve generations ago. This has a bearing on domestic organization. In a long-established township or village the maximal matrilineages which make up the community are clearly defined corporate groups, which until recently occupied distinct wards in the township. The most important political offices are vested in such lineages; and the most significant events in a person's life, marriage and death, both require the participation of the lineage acting as a corporate body.

The difference in internal cohesion between the two townships has historical roots. Owing perhaps to its relative geographical isolation until recently, Agogo has escaped severe disturbance from outside for several generations. Asokore has been less fortunate. In 1870–1 Asokore was involved in a rebellion against the King of Ashanti. Its failure resulted in the dispersal of the people of the chiefdom. Some were sold into captivity, many fled across the frontiers into what is now the state of New Juaben, most of them scattered into the forests of the chiefdom. Some years later a number of the survivors returned to rebuild Asokore township. Thus today we find at Asokore only rump lineages, segments of maximal lineages whose members are scattered far and wide, whereas at Agogo we find that the majority of the members of any lineage have their permanent homes in the township.[1]

[1] More detailed information about the social composition of Agogo is provided in Fortes 1954.

THE DWELLING GROUP

The Ashanti live in rectangular houses clearly separated from one another. A dwelling-house (*fie*) is occupied by what I shall call a dwelling group. Ashanti domestic organization, nowadays, is very elastic. Not only do dwelling groups vary in composition at a given time, but their membership fluctuates from month to month as people move to and fro for farming or trade. Moreover, men and women who normally reside elsewhere pay frequent visits to the townships where their matrilineages are domiciled. They are then counted as full members of the matrilineage and domestic group. Citizenship is conferred by the fact of matrilineal descent, and an Ashanti always remains a subject of the chiefdom to which his matrilineage belongs. Indeed, women are obliged by traditional custom to return to their mothers' homes to bear their babies, and they generally do so still. To establish the norms regulating association in domestic groups it is important therefore to examine them at a time of the year when movement in and out of the townships is least. The data used were obtained at such a time. The records include a complete census of Asokore township and a census of one in four of the households of Agogo.

The most striking feature of Ashanti domestic life appears vividly in one of the common sights in any village or township. As night falls young boys and girls can be seen hurrying in all directions carrying large pots of cooked food. One can often see food being carried out of a house and a few minutes later an almost equal amount of food being carried into it. The food is being taken by the children from the houses in which their mothers reside to those in which their fathers live. Thus one learns that husband and wife often belong to different domestic groups, the children perhaps sleeping in their mothers' houses and eating with their fathers. But inquiry shows that husband, wife, and children sometimes occupy the same dwelling. Frequently, however, the domestic group appears to consist of women only or of a miscellaneous assortment of men, women, and children.

Genealogical analysis shows that the norm is for the dwelling group to be a single kin group, that is, one in which the members are all connected with one another by kinship or marriage.

Where more than one kin group occupies a dwelling this is due to parts of houses being let to non-relatives, a common practice in congested areas. At Agogo 90 per cent. of all dwelling-houses are occupied by single kin groups; at Asokore the proportion is 69 per cent., the remaining 31 per cent. of the houses having tenant families as well as the owners. This is one of many effects of the close urban contacts of Asokore.

The dwelling group does not, as a rule, have a common food-supply nor do its members pool their incomes for the common support. But the norm is for the dwelling group to consist of a single household in the social sense, that is, a group in which the rule holds that food and assistance are freely asked and given between members. In Ashanti every adult is expected to earn the major part of his or her livelihood, and this means that mutual aid occurs as an obligation of kinship, not as a part of domestic organization. It extends to kinsfolk outside the domestic unit, on the same terms as to those within it. Though the dwelling group is the most clear-cut domestic unit, Ashanti include other near kin, notably lineage kin, in their concept of the household. Hence the term *fie*, house, is used not only for the dwelling group but also for a segment of the maximal matrilineage.

GENERATION DEPTH

An important aspect of domestic organization is the normal duration of the unit. It is an index of the process of growth by which the physical and social replacement of one generation by the next is assured and depends on the functions performed by the domestic unit in the rearing and education of children. In our society the domestic family seldom lasts more than one generation or, therefore, includes more than two successive generations. In both Agogo and Asokore, 45 per cent. of all households were of this kind; 55 per cent. included three or more generations, but only a small minority had more than three generations. No doubt this is influenced by a low rate of survival till old age. The net result, in any case, is that the domestic group tends to last only as long as its members are bound to one another by direct, first-order[1] kinship ties. This is correlated with the fact

---

[1] That is, relationships of the order of parent and child or sibling and sibling. (See Fortes 1950, 1954.)

that continuity in the social structure, so far as it depends on the recognition of genealogical connections, is maintained by the lineage system.

THE HOUSEHOLD HEAD

Genealogical analysis shows that the position of household head (*fie panin*) is the key to the most important features of domestic structure. Both men and women occupy this position, and Ashanti maintain that there is complete equality between them in this respect. They say also that the head of a dwelling group is normally the most senior by age, generation, or status of the members. This follows from the rule that seniority carries authority and commands respect. But it is also a basic rule in Ashanti social and political organization that seniority alone does not automatically confer positions of authority. A household head must have the necessary personal qualities to command the respect and adherence of the group and, in particular, he must have the means to maintain his station. A person of junior genealogical or legal status can also be a household head. Much depends on how the position is acquired; and the main factor in this is possession of the dwelling-house. This may be acquired by inheritance, bequest, or gift; but in very many cases it is the result of personal economic achievement.[1] An independent household is very often set up by the head's building or buying a dwelling-house. This may take place for occupational reasons, on marriage, or simply because the person has the money. Thus a young man may be *de facto* owner of the house, but his mother may live with him and be recognized as titular head. The head's position is primarily a status in relation to which co-operation, harmony, and cohesion are maintained in the group. He has no control over the other adult members in economic affairs, and can exercise rights over their persons or property only with the consent of the lineage segment concerned.

Examination of the field data shows that practice conforms to the ideal, the norm being for the household head to belong to the oldest generation in the group. In two-generation groups this is invariably the case; when three or more generations are

[1] Few dwelling-houses in townships like Agogo and Asokore are more than thirty years old.

represented in the domestic unit, we find that this rule holds for 91 per cent. of cases at Asokore and for 80 per cent. of the cases at Agogo; and it is of interest to note that there are no differences in this respect between male and female heads.

At Asokore 60 per cent. of all household heads are men, at Agogo only 53 per cent. This difference may be due to sampling variations, but it is also consistent with differences in household structure which will be discussed presently and which are correlated with differences in the total social structure of the two places.

The following table throws further light on the position of the household head.

TABLE I. Age Distribution of Household Heads, in Years[1]

		20–4	25–30	31–40	41–50	Over 51	Totals
Male heads							
Asokore	No.	1	10	12	13	25	61
,,	%	2	16	20	21	41	100
Male heads							
Agogo	No.	..	6	19	14	25	64
,,	%	..	9	30	22	39	100
Female heads							
Asokore	No.	2	5	6	6	18	37
,,	%	5	14	16	16	49	100
Female heads							
Agogo	No.	1	5	12	11	28	57
,,	%	2	9	21	19	49	100

Too much importance must not be attached to the detailed figures, as the samples are small. The broad picture, however, is clear. In both places just over 60 per cent. of the male heads and rather more than 65 per cent. of the female heads are over 41. By contrast it is almost impossible for a man or a woman to become a head before the age of 25. At Agogo, in fact, it is rare for this to be possible before the age of 30. At Asokore the position seems attainable at an earlier age than at Agogo.

[1] Special care was taken to secure age estimates correct to within about three years.

The following table lends point to this conclusion.

TABLE 2. Percentage of each Age-group in the Population who
are Household Heads

Age-group	20–4	25–30	31–40	41–50	Over 51	Total
Males						
Asokore	5	40	41	65	93	50
Agogo	0	21	38	47	57	35
Females						
Asokore	5	13	14	21	64	20
Agogo	2	10	20	19	42	19

From this table it appears that men have twice as much chance
as women of becoming household heads. It seems also that
Asokore men achieve the position more readily than Agogo
men at all ages. The latter are content to remain members of older
relatives' households to a greater measure and for longer periods
than the former. Other evidence confirms this and indicates that
it is due to the greater strength of lineage ties at Agogo than at
Asokore. There is little difference among the women of the two
places except in the oldest age-group. It seems that four out of five
women remain dependent members of households for most of their
lives and that they only become free to set up their own households after they have passed the child-bearing years. These figures
give a measure of the influence of differences in social and economic
opportunities associated with sex. They also show that a factor of
growth, correlated with age, influences domestic structure.

HEADSHIP AND SOCIAL MATURATION

The processes symbolized by the correlation of headship of
households with sex and age cannot be discovered by statistical
means. They are processes of social maturation in the individual
taking place in relation to the changing structure of the domestic
unit. Some of the variables involved have general incidence,
others are very restricted. For example, Ashanti lay great stress
on the economic circumstances which enable a person to set up
an independent household. Indeed every Ashanti, man or woman,
aspires to have his or her own house. But it is only in periods of

exceptional prosperity that a man can achieve the means for this before middle age. Though women in theory have the same economic opportunities and legal status as men, in practice their freedom is less. Thus their civic status is, in fact, subject to limitations from which men are free. Except for queen mothers, women cannot hold political office. Women have the right to express their views freely in lineage affairs; but they cannot be elected to the headship of a lineage.[1] Women occasionally become wealthy and many own more property than their menfolk; but in general the responsibilities of motherhood while the children are young are a handicap to economic achievement. So it is more common for a woman to become the head of a household as a result of her son's building a house for her than by her own efforts. In fact, as will be clearer presently, a woman becomes head of a household primarily in virtue of maternal kinship ties. It is rare for a woman without offspring to be able to gather a household around her.

Among many motives influencing men to set up their own household are Ashanti ideas of dignity. A young man does not mind residing with an older brother or his father or his mother's brother. A man of middle age would regard this as unbecoming. Of course no man of means would be content to live as a dependent member of a brother's or mother's brother's household. He would be expected to take responsibility for young members of his mother's lineage segment, and in particular, to have a house where some or all could live. One of the strongest motives is the desire, among both men and women, for domestic independence of the effective minimal lineage consisting of the children and daughters' children of one woman (as is apparent from Table 7a).

For both men and women marriage is a turning-point in social maturation. In Ashanti only men and women suffering from serious physical or mental infirmity remain unmarried. The mean age of first marriage is between 25 and 30 for men and between 15 and 20 for women. Until marriage both men and women are regarded as not fully adult and accept their position of dependence on senior relatives. After marriage the ideal is for a

[1] Women are barred from holding political office on account of the menstrual taboos. Menstruant women are forbidden contact with sacred objects; and political office, like headship of a lineage, is connected with the ancestral stools and involves religious duties on occasions such as the Adae ceremonies. See Rattray (1927, ch. 7).

man to have his own home and to have his wife and children
living with him. But this ideal is less often realized than Ashanti
believe. Even though it is nowadays buttressed by Christian
teaching, which has a wide influence, the ideal of the 'patrilocal'
family as the normal domestic group has not wholly asserted
itself. The bonds of matrilineal kinship work too strongly against
it.

FORMS OF DOMESTIC UNIT

Genealogical data from Asokore and Agogo (and these are
confirmed by data from other rural areas) show that three forms
of domestic unit can apparently be distinguished[1] by inspection,
as follows:

A. Households grouped around a husband and wife. In the
   simplest case this corresponds to the elementary family
   consisting of a man, his wife, and their children; but other
   kinsfolk may be included in the group.
B. Households grouped around an effective minimal matrilineage or part of it, such as a woman and her sister or
   daughters, or a man and his sister or sister's son.
C. Households made up of combinations of the previous
   types, e.g. a household consisting of a man and his wife and
   children as well as his sister's children.

In addition there are generally in every village a few miscellaneous dwelling groups consisting either of remnants of
larger households or of an assortment of kinsfolk living together
for convenience.

This classification follows the usual ethnographic practice.
Family histories show that these three 'types' are not a result of
culture contact but were certainly common fifty years ago. The
classification can be simplified if it is reformulated by reference to
the focal position of the head in the structure of the domestic
group. We see, then, that households with male heads can be:
(a) 'patrilocal', that is, made up of a man and his dependants
by marriage and paternity; or (b) 'avunculo-local', that is,

---

[1] The effect of polygyny is ignored in this analysis, as it is not a significant
factor. Polygynists sometimes have one or more of their wives living with them
in a 'patrilocal' household, though it is generally agreed that it is better for the
wives to live apart in order to avoid jealousy and friction.

consisting of a man and his dependants by matrilineal kinship; or
(c) a mixture of the two. Similarly, households with female
heads can be: (a) 'matrilocal', that is, made up of a woman and
her dependants by marriage and motherhood; or (b) 'matrilineal',
consisting of a woman and her dependants by matrilineal kinship;
or (c) a combination of the two. This way of stating the facts has
the merit of drawing attention to the factors that appear to
underlie the variations in type. One factor is the attainment of
headship of a household, another is marriage and parenthood, a
third is matrilineal kinship; and it can be suggested that the type
of domestic unit found in a particular case is a result of the balance
struck between the obligations of marriage and parenthood
on the one hand and those due to matrilineal kin on the other.
But we can go farther if we resort to more detailed numerical
analysis.

KINSHIP IN THE DOMESTIC UNIT

The problem is: How are the members of a household normally
connected with the head of the group? By Ashanti law and custom,
one's mother's home – that is, by definition, the place where one's
matrilineage is domiciled – is one's true home. A person therefore
has a moral, if not a legal, right to house-room in the house of any
member of his matrilineage. This is associated with the corporate
status of the lineage, and the closer the lineage tie the more
precisely legal is this right. But Ashanti say, also, that a person can
claim house-room in the house of a patrilateral kinsman and that
it is morally unforgivable[1] to refuse him even though it is legally
permissible to do so. While, however, lineage ties are recognized
in relation to common ascendants up to ten or twelve generations
back, patrilateral ties are rarely known or recognized (among
commoners) in relation to ascendants more than four generations
back.

   In theory, then, a dwelling group can include a wide range
of kinsfolk. We find, in fact, that if the population we are dealing
with is classified in accordance with their exact genealogical
relationships with the heads of their respective households, they

[1] It would be an affront to the spirits of common patrilateral ancestors; and
though Christian Ashanti would not put it this way they nevertheless accept the
principle.

fall into the following numbers of groups:

Asokore   males: 35 groups   Agogo   males: 42 groups
   ,,   females: 30 ,,   ,,   females: 47   ,,

If these are rearranged according to native classificatory categories, they fall into ten groups. The range includes lineal kin from mother's mother to daughter's daughter's child of the household head, and a great variety of collateral kin. But many of the latter occur only once or twice in the combined sample of 1,676 (715 males and 961 females, exclusive of heads). In fact, about 90 per cent. of the members of households are close kin connected with the head by one (e.g. sister's son) or at most two (e.g. mother's sister's son) intermediate links.

The following table elucidates this argument. To simplify comparison the frequencies in each cell are given as approximate percentages of the total sample for each township.

TABLE 3. Classification of Members of Households by Kinship with Head

| | Members living in households with | | | | Totals living with | |
| | Male heads | | Female heads | | | |
	Lineage kin	Non-lineage kin	Lineage kin	Non-lineage kin	Male heads	Female heads
	As. Ag.	As. Ag.	As. Ag.	As. Ag.	As. Ag.	As. Ag.
Male members	9 12	20 10	16 15	2 4	28 22	17 19
Female members	7 16	26 17	18 22	3 5	33 32	22 27
Totals	16 28	46 27	34 37	5 9	61 54	39 46

As. = Asokore (percentages of total sample of 711 = 324 males plus 387 females)
Ag. = Agogo (,,   ,, ,,   ,,   ,, 965 = 391 ,,   ,, 574 ,,)

Non-lineage kin, in the above table, includes all members of households who are related to the head by marriage or paternity. Sons and daughters living with both parents in 'patrilocal' households, as well as their offspring, are included in this category. 'Lineage kin' includes all matrilineal kinsfolk.

It must be stressed again that too much importance must not be attached to the actual figures. The general picture is, however, clear, and it is significant that there is good agreement between the two samples. We see that a substantially larger proportion of people of both sexes live in households under male heads than in households with female heads. This is consistent with the greater relative incidence of male heads in the population as well as with the more favourable and influential status of men in economic and legal matters. A point of special interest, though, is the very small difference in favour of male heads at Agogo by contrast with their large excess at Asokore. It seems that the balance of authority between the sexes in the domestic sphere is almost equal at Agogo, whereas men have the lead at Asokore. This is consistent with the greater strength of lineage solidarity at Agogo, to which attention was drawn earlier.

A striking feature, in both places, is the very small proportion of non-lineage kin living under the domestic jurisdiction of female heads. Genealogical analysis shows, in fact, that households under female heads are segments of a matrilineage (see Tables 7a and 7b). This is in keeping with Ashanti kinship and jural values. For a man to live in his wife's house is considered to be contemptible and only five cases, all the result of peculiar circumstances, occur in our total sample. But it is no disgrace for a man to live in his mother's or his uterine sister's house. Through the whole gamut of social relations in Ashanti there is no tie so fundamental and so strong as that of mother and child. Though it has the sanction of the whole of customary law, on account of the dominance of the principle of matrilineal descent, Ashanti think of it as an ultimate and irreducible moral and psychological fact which needs no sanctions.[1] A woman and her children constitute the indissoluble core of Ashanti social organization. A woman's primary responsibility is for her own children; her devotion to them outweighs everything else in life; and they, in turn, never question her position of privilege and respect.

An important structural principle springs from this. The members of the matricentral group, the mother as well as her children, seek to assert its autonomy wherever possible, and to maintain its unity as long as the social relations created within it survive. Full autonomy is reached when the mother, if she lives long enough, or one of the sons, or, less commonly, daughters, becomes head of a household. Inevitably, therefore, a household under a female head must be a matrilineage segment. It follows also that the majority of the non-lineage members of such a

[1] These ideas are graphically expressed in many Ashanti proverbs and maxims.

household must be, as analysis shows them to be, the children of brothers or sons of the head, whose wives are living with them (see Tables 6 and 7).

A matricentral household maintains its unity as long as the head survives, which may be till the third or at most fourth generation. When she dies her children may wish to separate into independent units. At Asokore this tendency is marked owing to the instability of the lineage organization and the opportunities for individual enterprise offered by the nearby city. At Agogo the social structure and the economic environment favour matrilineal solidarity, and the group often lasts as long as any of the original head's children survive. Thus we find relatively few cases of adult brothers and sisters residing together at Asokore, whereas it is quite common at Agogo, as can be seen from Table 4.

To sum up, the matricentral household is the farthest extension in time of the elementary matricentral cell of mother and children, its duration being to a large degree conditioned by the possibility of its members having direct contact with, and experience of dependence on the original head.

Households under male heads present a different picture. At Asokore they have more than twice as many non-lineage as lineage kin among the male members and nearly four times as many among the female members. At Agogo the two classes of kin are about equal, as might be expected from what we know of the greater strength of lineage ties there.

The following tables help to elucidate the picture.

TABLE 4. Classification of Members of Households with
Male Heads (excluding Heads)

A. MALE MEMBERS

	Asokore		Agogo	
	Number	Per cent	Number	Per cent
I. *Lineage kin*				
Full brothers	6	3	15	7
Full sisters' sons	38	19	56	27
Classificatory sisters' sons	3	1.5	..	.. (a)
Sisters' daughters' sons	7	3.5	30	14
Other matrilineal kin	9	4.5	16	8 (b)
Total I	63	31.5	117	56

	Asokore Number	Asokore Per cent	Agogo Number	Agogo Per cent
II. Non-lineage kin				
Own sons	94	47	61	29
Daughters' sons	24	12	10	5
Other patrilateral kin	16	8	13	6
Miscellaneous kin	4	2	10	5
Total II	138	69	94	45
Totals I and II	201	100	211	100

B. FEMALE MEMBERS

	Asokore Number	Asokore Per cent		Agogo Number	Agogo Per cent	
I. Lineage kin						
Full sisters	14	9 (c)	6 (d)	24	9 (c)	8 (d)
Full sisters' daughters	26	16	11	71	26	23
Classificatory sisters' daughters	..	..	..	18	7	6
Sisters' daughters' daughters	1	1	0.4	15	6	5
				{ 7	{ 3	{ 2 (e)
Other matrilineal kin	8	5	3 (f)	{ 38	{ 14	{ 12 (f)
Total I	49	31	20	173	65	56
II. Non-lineage kin						
Daughters	83	52	36	64	24	20
Daughters' daughters	23	15	10	17	6	5
Other patrilineal kin	3	2	1	15	6	5
Total II	109	69	47	96	36	30
III. Spouses and affines						
Wives	71		31	46		15
Wives' kin	5		2	..		..
Total III	76		33	46		15
Totals I, II and III	234	100	100	315	100	100

(a) Mainly mothers' sisters' daughters' sons.
(b) Includes a number of mothers' brothers.
(c) Percentages exclusive of wives and their kin.
(d) Percentages including wives and their kin.
(e) Mostly more distant than mothers' mothers' descendants.
(f) About half are mothers and mothers' sisters.

These tables include males and females of all ages from infancy to old age. It is worth noting that the majority belong to the next succeeding (i.e. filial) generation after the heads (*cf.* also Tables 6 and 7). But what is most significant is the apparent contrast between Asokore and Agogo. The striking preponderance of sisters' uterine grandchildren, sisters' children, and mothers at Agogo as compared with Asokore testifies to the strength of the matricentral unit over a long stretch of time. This appears also from the figures which show that in both places collateral kin beyond the range of the lineage sprung from the head's own mother are rarer at Asokore than at Agogo.

The key relationships in the household under male heads are obvious (see italicized figures). They are the relations of husband and wife, brother and sister, father and child, uncle and sister's child. Conjugal and paternal ties must tend to go together, as a woman living with her husband very likely has her young children with her. But a man may have his child in his household though its mother lives apart. This applies even more to his sister's child.

At Asokore there is preponderant stress, in these households, on the conjugal and parental relationships; at Agogo there is practically equal balance between them and the tie with sisters and their children. But in both places (as throughout Ashanti) the structure of the household under a male head is the resultant of the balance struck between these two classes of social bonds.

FORM AND STRUCTURE

One conclusion is obvious. We are dealing not with two 'types' or 'forms' of domestic organization but with variations of a single 'form' arising out of quantitative differences in the relations between the parts that make up the structure. We can imagine a scale varying from perfect 'patrilocality' at one end to perfect 'avunculo-locality' at the other. The normal or modal Agogo household would come about half-way between the extremes, and the modal Asokore household about half-way between the centre and the 'patrilocal' end. Individual households are scattered all along the scale; and over a stretch of time a particular household may change its position through the loss of some kinsfolk and the accession of others. In simpler terms, the norm for Asokore is that when a man is head of a household he has his wife and children living with him even if this means excluding his maternal kin, whereas at Agogo he compromises so that neither side is favoured at the expense of the other.

The dynamics of this process can be understood in terms of Ashanti kinship values. As in other matrilineal societies there is a polar relationship between the ties of matrilineal kinship and those of marriage and parenthood. This is manifested in Ashanti legal and religious institutions as well as in emotional attitudes and individual prejudices. Ashanti discuss the subject interminably, stressing especially the inevitability of conflicting loyalties. For a woman the conflict turns on the difficulties of reconciling attachment to her mother with duty to her husband. A man feels it most in relation to his own children and his sister's children. For both it is epitomized in the contrast between the mutual trust and loyalty of brother and sister and the notorious hazards of marriage.

Matrilineal descent has a compelling influence on conduct, because it is the basis of the lineage and therefore determines political allegiance, rights of succession and inheritance, the regulation and validation of marriage, and corporate support in such crises as death and, nowadays, economic distress. A man is legally obliged to consider the interests of his sister's children because he is their legal guardian and they are his potential heirs. He is morally impelled to care for them because they ensure the continuation of the lineage segment sprung from his mother. But exceptional importance is also attached to paternity – though patriliny is not legally recognized – in law, in religion, and in personal relations. It is considered to be a father's duty to bring up and equip his children for life. His love for them and pride in them are often stronger than his legally enjoined responsibility for his sister's children. Ashanti say that the traditional compromise is for a man to have his children living with him till adolescence and then to let them go to their mother's brother. They describe cross-cousin marriage as a device by which men try to unite their love of their children with their loyalty to their maternal kin. These norms and attitudes have their counterparts in the norms of filial duty (e.g. children have to provide their father's coffin), respect, and affection, the spontaneity of which is often contrasted with the obligatory duty and respect towards the uncle. Christianity and modern changes have aggravated but not created these tensions.[1] It is of great significance that there is not a single case, in either township, of a man and his wife's brother sharing a dwelling. This structural arrangement prevents intolerable conflicts of loyalties. In the domestic environment a child is never confronted with both men of the parental generation who have authority over him.

THE TIME FACTOR: WIVES

I have emphasized that Ashanti domestic structure changes over time in a manner analogous to growth. Does this influence the kind of compromise reflected in domestic organization at a particular time? It may be, for instance, that while her children are young a woman prefers to reside with her husband, but that when they grow up she prefers her own matricentral household. The residence of wives is a key issue in this. If none of them resides with her husband there can be no 'patrilocal' household. Since social maturation is correlated with age, an analysis of the connection between age and residence of wives may throw light on the problem. This is done in the following table:

TABLE 5. Age, Residence, and Marital Status of Women

Age-group (years)		Married				Not married (including widows and divorced women)		Totals	
		Living with husband		Not living with husband (a)					
		As.	Ag.	As.	Ag.	As.	Ag.	As.	Ag.
16–20:	No.	6	8	14	30	32	23	52	61
	%	12	13	27	49	61	38	100	100
21–25:	No.	20	10	13	39	11	15	44	64
	%	45	16	30	61	25	23	100	100
26–30:	No.	18	6	15	33	5	12	38	51
	%	47	12	40	65	13	22	100	100
31–40:	No.	18	14	17	31	10	15	45	60
	%	40	23	38	52	22	25	100	100
41–50:	No.	9	15	11	19	9	23	29	57
	%	31	26	38	33	31	40	100	100
Over 50:	No.	9	7	4	16	16	43	28	66
	%	32	11	14	24	57	65	100	100
Totals		80	60	74	168	83	131	236	359

(Grouped bracket figures: 21–40 Living with husband As. 56, Ag. 30 (44/17); Not living with husband As. 45, Ag. 103 (36/59); Not married As. 26, Totals 42 (20/24). 41–Over 50 Living with husband As. 18, Ag. 19; Not living with husband 15/35, 26/28; Not married 25, Totals 66/53.)

As. = Asokore.   Ag. = Agogo.   (a) The majority living with close maternal kin.

[1] As Ashanti themselves point out, and as can be seen from Rattray's penetrating description (1929, chs. 1–7).

Sampling and other errors make it necessary to treat the specific frequencies with reserve. Nevertheless the general trend brings out clearly the effect of social maturation. During the first two or three years of wifehood the great majority of young wives continue to reside with their own kin. Young and inexperienced, they cling to their mothers. As they advance in maturity the pull of conjugal ties increases and reaches its maximum at the peak of the child-bearing years, in the thirties. By this time they have perhaps three children, for whom it is an advantage to be under their father's care. Finally, when their child-bearing years are over and their children are grown up, the desire to establish their own households becomes strong in many women, the more so if they are widowed or divorced.

But it must be remembered that attachment to the maternal home, the pull of conjugal ties, and the desire for independence are all active throughout a woman's and indeed a man's life. Local norms arising out of the total social structure of the community are important. The bias at Asokore is in favour of conjugal ties, but so strong are the ties of matrilineal kinship that nearly half of all the married women, even at the peak of the child-bearing years, prefer to live with their own kin. At Agogo, in keeping with what we have already learnt, the pull of matrilineal kinship is three times as strong as that of marriage, at the peak of child-bearing. Indeed, the norm at Agogo is so markedly against a wife's living with her husband that special circumstances must be looked for to explain the exceptions. Cross-cousin marriage is one such factor. Among 41 Agogo wives living with their husbands, 17, or nearly half, were found to be their husbands' cross-cousins. These data support the Ashanti rationalization of cross-cousin marriage. Investigation also shows that nearly half of the Asokore women who live separately from their husbands reside with their mothers, and if those who live with both parents are added the proportion comes to 65 per cent. The corresponding figures for Agogo are 35 and 43 per cent., a further 32 per cent. living with collateral lineage kin, which is rare at Asokore. At Asokore, as ethnographic inquiry confirms, it appears that it is specifically the attachment of a woman to her mother rather than, as at Agogo, more inclusive lineage solidarity, that deters her from residing with her husband.

TABLE 6a. Age and Relationship to Head of Filial Members
of Households under Male Heads

| | | | A. Patrilineal kin | | | | B. Matrilineal kin | | | | | | | Total | | Total | |
| | | | Brothers' and sons' (b) | | Daughters' | | Sisters' (c) | | Sisters' daughters' | | Other matrilineal kin | | | | | | |
Age-group	Sons (a)	Daughters (a)	S	D	S	D	S	D	S	D	M	F	A	B	M	F
Asokore																
Under 15	73	62	7	0	18	22	26	13	7	1	4	3	182	54	135	101
16–25	17	16	5	2	6	0	11	7	0	0	2	1	46	21	41	26
26–40	4	5	3	1	0	1	3	4	0	0	2	2	14	11	12	13
Over 40	0	0	1	0	0	0	1	2	0	0	1	2	1	6	3	4
Total As.	94	83	16	3	24	23	41	26	7	1	9	8	243	92	191	144
Agogo																
Under 15	43	42	5	12	7	14	27	36	30	30	7	4	123	134	119	138
16–25	12	14	6	2	3	3	13	19	0	2	3	0	40	37	37	40
26–40	6	6	1	1	0	0	15	15	0	1	4	3	14	38	26	26
Over 40	0	2	1	0	0	0	1	1	0	0	2	0	3	4	4	3
Total Ag.	61	64	13	15	10	17	56	71	30	33	16	7	180	213	186	207

(a) Approximately 80 per cent. at Asokore and 70 per cent. at Agogo living with both parents.
(b) Including a few classificatory kin in these classes.
(c) Including a few classificatory sisters' sons.
As. = Asokore. Ag. = Agogo. S. = Sons. D. = Daughters. M. = Males.
F. = Females.

THE TIME FACTOR: CHILDREN

From the foregoing analysis it appears that the possession and the ages of children are also significant factors in domestic structure. This problem is examined in tables 6a, 6b, 7a and 7b, which show the age-distribution of all members of households who stand in a filial relationship to the head. They include 98 per cent. of all children under 15 in the two samples.

In both Agogo and Asokore between 65 and 70 per cent. of the filial members of households under male heads are children under 15. For households under female heads the proportions are: 73 per cent. of the males and 55 per cent. of the females at Asokore; 70 per cent. of the males and 55 per cent. of the females at Agogo. Clearly the local bias in domestic structure does not affect the

TABLE 6b. Proportions of Own Children and Sisters'
Children living in Households of Male Heads

							Total	
Age-group (years)	Sons	Daughters	Sisters' sons	Sisters' daughters	Other males	Other females	Males	Females
ASOKORE								
Under 15:								
No.	73	62	26	13	36	26	135	101
%	54	62	19	13	27	25	100	100
Over 15:								
No.	21	21	15	13	20	9	56	43
%	38	50	27	30	35	20	100	100
AGOGO								
Under 15:								
No.	43	42	27	36	49	60	119	138
%	36	30	23	26	41	44	100	100
Over 15:								
No.	18	22	29	35	20	12	67	69
%	27	32	43	57	30	11	100	100

% shows the percentage of sons, sisters' sons, and other males in relation to all males of the age-group, and the same for females of each class. Thus at Asokore sons comprise 54 per cent. of all males under 15 and 38 per cent. of all males over 15; daughters are 62 per cent. of all females under 15 and 50 per cent of all females over 15.

TABLE 7a. Relationship to Head of Filial Members living under Female Heads

Age-group (years)	A. Patri-lateral kin Brothers' and sons'		Sons (a)	Daughters (a)	B. Matrilineal kin Sisters'		Daughters' and sisters' daughters'		Other matri-lineal kin		Totals		Totals	
	S	D			S	D	S	D	M	F	A	B	M	F
Asokore														
Under 15	7	5	32	28	1	2	45	41	0	0	12	149	85	76
16–25	2	3	10	21	1	1	8	13	0	0	5	54	21	38
26–40	0	0	9	15	0	0	0	4	0	0	0	28	9	19
Over 40	0	0	1	4	0	0	0	0	0	0	0	5	1	4
Total As.	9	8	52	68	2	3	53	58	0	0	17	236	116	137
Agogo														
Under 15	5	10	36	35	13	10	59	75	2	1	15	231	115	131
16–25	1	5	14	34	1	2	7	11	1	7	6	77	24	59
26–40	0	2	8	24	3	2	0	4	6	5	2	52	17	37
Over 40	0	0	4	8	0	0	0	0	4	5	0	21	8	13
Total Ag.	6	17	62	101	17	14	66	90	13	18	23	381	164	240

(a) Approximately 85 per cent. at both places living with mother only, father living separately.
As. = Asokore. Ag. = Agogo. S. = Sons. D. = Daughters.

TABLE 7*b*. Proportions of Own Children living in
Households of Female Heads

Age-group (years)	ASOKORE				Total	
	Sons	Daughters	Other males	Other females	Males	Females
Under 15: No.	32	28	53	48	85	76
%	38	37	62	63	100	100
Over 15: No.	20	40	11	21	31	61
%	65	66	35	34	100	100
	AGOGO					
Under 15: No.	36	35	79	96	115	131
%	31	27	69	73	100	100
Over 15: No.	26	66	23	43	49	109
%	53	60	47	40	100	100

rule that the majority of dependent members of a household are
pre-adolescent children.

In households under male heads, moreover, it is the relationship
to the head that counts and sex is not significant. In households
under female heads, on the other hand, female dependants tend
to stay on in adulthood to the same extent as in childhood. This
is in keeping with what we have already learnt of the cohesion
of the matricentral unit through time. The household under a
male head disperses gradually as the children grow up.

The crucial question is the respective place of own children
and matrilineal dependants in household structure at different
stages of social maturation. As these two classes of kin are merged
into one in households under female heads the responsibility of
the head is not divided between the potentially rival claims of
own children and sister's children. A woman's brothers' or sons'
children hardly come into the picture, as she has no jural status
in relation to them. Hence a household under a woman head
consists almost entirely of her matrilineal kin. As she is generally
elderly (*cf.* Table 1) the majority of her adult filial dependants
must be her own children, as Tables 7*a* and 7*b* show, while those
under 15 are bound to consist mainly of her (and her sisters')
uterine grandchildren. These tables confirm the view that an
adult will only stay in a household under a female head if she is

his or her own mother or, less commonly, mother's sister; and a man will generally try to establish his own household rather than stay on as a dependant in a woman's household. Hence we find that only 1 in 10 of the male dependants in households with female heads is over 25 at Asokore, the corresponding figure for Agogo being about 1 in 7. Thus in general terms the norm is that the structure of a woman's household is based primarily on her children, especially her daughters, and it is very little affected by their social maturation. This norm, moreover, holds independently of the bias arising from the relative weight of conjugal ties and matrilineal ties by local standards. Conjugal ties are almost wholly excluded.

Households under male heads appear to present a different emphasis. Patrilateral dependants are as significant as matrilineal dependants, in fact more so at Asokore. But the important point is that these patrilineal adherents consist almost entirely of the children of the head. Their presence in the household is, however, markedly influenced by factors that are marginal in the structure of women's households; firstly, by the local bias in regard to conjugal and paternal ties; secondly, by the factor of social maturation; and thirdly, by the divided jural role of the head as father and uncle.

The local bias comes out in the contrast between the figures for Asokore and for Agogo. In keeping with our previous conclusions it is clear that a man's household at Asokore is much more likely to be based on his own than on his sisters' children, whereas at Agogo the chances of its being based on own children and on sisters' children are about equal. The decisive element, however, is not the desire of a man to have his children with him, though this is very influential. It is the fact of his wife's residing with him. This can be inferred from the observation that the great majority of children who live with their father do so because their mothers are living there. This inference is strengthened by the large proportion of sons and daughters in the under-15 age-group at Asokore where, as we know, the incidence of wives living with their husbands is very high (Table 4). The basic rule is that pre-adolescent children reside where their mothers live. The majority of pre-adolescent sisters' children also live under their mothers' immediate care. It is significant that a large proportion of the 16–25-year-old girls in this sample are daughters of the household heads, and the evidence previously brought forward shows that young women of this age stay either with their mothers or with their husbands. Thus the rule just stated works in favour of the 'patrilocal' household where conjugal ties are stressed and in favour of the matrilineal household where matrilineal ties are stressed.

But perhaps the most interesting figures in these tables are those showing the proportions of sons and daughters to sisters' sons and daughters. There is a substantial drop in both places in the proportion of sons and daughters residing with their fathers after adolescence, and an increase – by nearly 100 per cent. at Agogo – in that of sisters' sons and daughters. The Ashanti generalization that children stay with their fathers till adolescence and then move to their mothers' brothers seems to be borne out. But account must also be taken of the desire of men to establish their own households. Sons are under greater pressure to do this than sisters' sons, since they have no rights of inheritance in their fathers' houses, whereas the latter have in their uncles' houses. The proportion of the former in the households of male heads therefore declines more rapidly than the proportion of the latter after adolescence, as can be seen from the tables. This applies to daughters and sisters' daughters also. A married woman whose mother is dead has no inducement to stay on in her father's house where her children have no legal rights, but she may well do so in her uncle's house where she and her children have a legal right to use and inherit house-room and to corporate support in their affairs.

CONCLUSION

I have endeavoured to show in this essay that elementary statistical analysis is indispensable for the elucidation of certain problems of social structure that arise in a society which is in process of becoming socially diversified. The futility of blanket-terms like 'patrilocality' and 'matrilocality' in this context is obvious. The use of numerical data has enabled us to see that Ashanti domestic organization is the result of the interaction of a number of fairly precisely defined factors operating both at a given time and over a stretch of time. Granted the dominance of the rule of matrilineal descent and the recognition of paternity in Ashanti law and values, the sex of the household head is the factor of first importance. It determines the main possibilities of the arrangement of kinsfolk in the domestic unit in relation to the polar values of 'matricentral' and 'patricentral' grouping. The other factors are the tendency to seek a compromise between the opposed ties of marriage and parenthood on the one hand and those of matrilineal kinship on the other; and the ideal that every mature person, especially a man, should have his own household. How these factors interact depends, among other things, on local social conditions and historical circumstances. The domestic arrangements I have described are only possible in the long-established, relatively stable capital townships of chiefdoms, where both spouses in every marriage are equally at home. In new villages the ordinary patrilocal household is more common.

This investigation arose out of a consideration of some of Radcliffe-Brown's most recent views on the nature of social structure; that it leads to conclusions not altogether in agreement with his generalizations is a tribute to their significance. Our investigation shows that elementary statistical procedures reduce apparently discrete 'types' or 'forms' of domestic organization in Ashanti to the differential effects of identical principles in varying local, social contexts. This makes an assessment of the factors underlying the 'norms' possible; and it also enables us to relate the 'norms' to one another and to the apparent 'types' of domestic organization by taking into account the effect of time as an index of growth. 'Structure' thus appears as an arrangement of parts brought about by the operation, through a period of time, of principles of social organization which have general validity in a particular society.

2

The Significance of Descent in
Tale Social Structure[1]

INTRODUCTORY

The Tallensi of the Northern Territories of the Gold Coast
furnish data of special interest for the study of comparative social
structure among the peoples of West Africa. Large as the ethnographic literature on West Africa is, it is singularly lacking in
analytical data concerning social structure. Some of the most
useful collections of ethnographic information on West African
peoples thus lack the foundation without which a coherent
picture of a society is impossible. Tables of kinship terms, enu-
merations of kinship usages, catalogues of marriage and inheritance customs, and such-like information are no more than the
raw materials for the construction of a systematic representation
of social structure. And very often the raw materials are not
sufficient. There are plenty of bricks but no mortar. The reasons
for such lacunae are obvious. A sympathetic amateur ethnographer can bring together material of inestimable value; but
without a good theoretical grounding in modern social anthropology the field worker will not look for, and even if he stumbles
across it, will not recognize the kind of material necessary for
an understanding of social structure. He must, first of all, have
the concept of a total social structure clearly in his mind; and he
must look for the connections, which are very often implicit,
by which ostensibly discrete processes and institutions are related
to one another in a meaningful pattern.

It is for this reason that we know very little about West
African lineage systems, though the more recent literature indicates quite clearly that the lineage is one of the most stable,

---

[1] Reprinted from *Africa*, XIV, 1943–4.

widespread, and fundamental units of social structure in West Africa.[1]

The special interest of the Tallensi lies in their exceptionally clear-cut lineage system and in the all-embracing sweep it has in their social system. The whole of Tale society is built up round the lineage system. It is the skeleton of their social structure, the bony framework which shapes their body politic; it guides their economic life and moulds their religious ideas and values. The high degree of stability and continuity of structure characteristic of Tale society rests on the lineage system. No aspect of Tale culture can be understood without a grasp of the principles that govern their lineage organization. It is the basis of their political and jural relations, as has been shown elsewhere,[2] and the supreme regulating factor in all corporate activities. Built up on and continuously recreated by the flow of kinship relations, the lineage system at the same time determines the channels in which kinship relations move.[3]

Apart from its interest as a specimen of what is probably the typical West African patrilineal lineage pattern,[4] the Tale lineage system is of general theoretical interest. There is first of all the ancient controversy about the family and the clan, or whatever the corporate unit may be that corresponds to the clan in the constituted framework of social structure. This centres round the

[1] The lineage is clearly the basic unit of jural and political organization among the Ibo (see Meek 1937) and the Yakö (see Forde 1938, pp. 311–38; and 1941) of south-eastern Nigeria. This appears to be the case also with the Gã of the Gold Coast (see Field 1940). It would appear to be the significant unit in respect of the laws of property, inheritance and succession, and legal responsibility among the matrilineal Akan peoples of the Gold Coast, as can be inferred from a careful reading of Rattray (1923 and 1929) side by side with Danquah (1928a and 1928b). Busia (1951) confirms this inference for Ashanti. In Dahomey, as Herskovits shows clearly (see Herskovits 1938), the lineage principle plays a very important part in every phase of social and economic life, and this is true also of the Yorubu (see Bascom 1944). The ethnographic literature of French West Africa shows that the lineage is probably the backbone of social structure among the Mole, Gurma, and Mandingo speaking tribes as is shown also in Rattray (1932) (see Labouret (1958) and Monteil (1924)).

[2] See Fortes (1940).

[3] A preliminary discussion of this subject was given in my paper on 'Kinship, Incest and Exogamy in the Northern Territories of the Gold Coast' (1936b). Both the terminology used in that paper and the data offered are superseded by the present paper. A full analysis of the Tale lineage system is given in Fortes (1945), excerpts from which have been used in the present paper.

[4] This is an hypothesis. But the Ibo, Yakö, Yoruba, and Dahomey data, referred to previously, make it seem highly probable.

question: Which comes first in time? Or, to transpose it into the current idiom of anthropological thought, which is source and which is consequence, the unilineal descent group or the family? An examination of this problem in the light of Tale material shows that it is at bottom an irrelevant one.[1]

The second theoretical problem on which Tale material sheds further light is that of the time factor in social structure. The concept of social structure postulates not only a rational and consistent relationship between the component parts of a social system, as Radcliffe-Brown constantly emphasizes,[2] but an enduring pattern in the system of social relations. This may be a fixed pattern or an evolving pattern, but the notion of a rational and consistent relationship between the component parts of a social system *in time* as well as *at a given time* is implicit in the concept of structure. It is, however, a difficult methodological task to incorporate the time dimension (which, be it noted, is not the same as chronology) into a synchronic analysis of the social structure of a community without written records. The difficulty is least in societies which have well-defined lineage systems, as Evans-Pritchard and Forde have demonstrated.[3] A lineage must be visualized both as a configuration at a given time and as a dynamic equilibrium in time. Tale material shows this very clearly. A Tale lineage always functions as a whole. Even when only part of a lineage emerges in corporate activity, the total lineage field is subliminally present and influential. A Tale lineage is a configuration of social processes, not a static grouping. It is continuously becoming internally differentiated. At any given time it incapsulates,[4] all its relevant past states. Owing to the fact that

[1] Forde (1941), pp. 114–16, discusses this question and says that the Yakö family is 'not the nuclear kinship unit from which other larger units (i.e. lineages) are built up'. This, as he maintains, is probably true of every society in which unilineal descent is recognized in the formation of corporate groups. But that is surely not the real issue, as can be seen from, e.g., Radcliffe-Brown's treatment of the subject (1930) or Firth's analysis (1936, 1957). The family can be thought of as the 'origin' of kinship relations in the same way as the heart would presumably be functionless without the whole circulatory system. Forde has, however, raised a theoretical problem of real importance for comparative studies of social structure, when he points out that the permanent and socially dominant structural machinery of kinship is the lineage and not the family, among the Yakö. This is true also of the Tallensi.   [2] e.g. Radcliffe-Brown (1935b).
[3] See Evans-Pritchard (1940a) and Forde (1938), *loc. cit.*
[4] This illuminating concept defining the relationship of past and present in historical process comes from Collingwood (1944).

non-agnates can never be absorbed into a lineage but can only
become attached to it as clearly differentiated elements, the
putative 'time depth' (Evans-Pritchard's term) and the contemporary span of a lineage are perfectly correlated.

In this paper, however, we are not concerned with the lineage
system of the Tallensi in all its aspects. We deal only with the
principles that govern lineage organization. These will be briefly
and formally stated in the form of a paradigm. Something of this
sort, though less systematic and abstract, is in the mind of every
well-informed Talǝŋ when he discusses the structure of his society
or takes part in public affairs. As the lineage is based on the rule
of patrilineal descent, we shall go on to discuss this rule and the
complementary principle of matrilineal kinship.

The Tallensi, it should be added,[1] though they number only
35,000, are typical of a considerable congeries of Mole-Dagbane
speaking peoples in the Northern Territories of the Gold Coast
and in the Upper Volta region of the French Ivory Coast.[2] All
these peoples are sedentary cultivators. The farming system,
based on mixed cropping, with millet and guinea corn as the
staples, and a loose scheme of rotational fallowing, enables them
to get only the barest livelihood out of the soil. The density of
population is relatively heavy, varying between 100 and 200 to the
square mile, throughout this area. The most cursory inspection
shows that the country of the Tallensi and their neighbours
has been continuously inhabited for a very considerable time.
The evidence of ancestral graves shows that the older Tale
settlements have certainly been occupied for at least eight to ten
generations by their present inhabitants. The settlements do not
form compact village units. Homesteads are scattered so as to
leave land for farming around each homestead. But the structural
basis of every settlement is a maximal lineage or an associated
group of maximal lineages, at least in the older parts of the Zua-
rungu District, of which Taleland forms a part. The Tallensi,
like most of the tribes of this region, have a markedly patrilineal
and patriarchal social structure.

[1] Further ethnographic details are given in my previously cited papers and in
Fortes, M. and S. L. (1936); see also Rattray (1932).
[2] These peoples occupy roughly the drainage area of the Volta River system
in British and adjacent French territory.

PARADIGM OF THE LINEAGE SYSTEM

A Tale lineage always functions as a system in which the behaviour of any part is regulated by its relationship to the whole, and the behaviour of the whole is determined by its component parts and the relations between them. This is the case whenever a lineage, or any part of one, emerges in social action. From the point of view of the Tallensi themselves, a lineage is an association of people of both sexes comprising all the known descendants by a known genealogy of a single known and named ancestor in an unbroken male line. From the sociologist's point of view, it is an association of people of both sexes comprising all the recognized descendants by an accepted genealogy of a single named ancestor in a putatively continuous male line. It is, in other words, a strictly unilineal, agnatic descent group.

Lineages vary in *span* proportionately to the number of generations accepted as having intervened between the living members and the founding ancestor from whom they trace their descent. The lineage of minimum span – the minimal lineage – consists of the children of one father; the lineage of maximum span – the maximal lineage – is the lineage of widest span to which any one of its members belongs. It consists of all the descendants in the male line of the remotest common patrilineal ancestor known to the members of the lineage. Whenever we speak of a lineage without indicating its span, we use the term to mean a lineage of any span and any order of segmentation. What is meant by order of segmentation is defined below. The number of antecedent generations reckoned to the point of convergent ascent varies slightly from one maximal lineage to another in Taleland. It is proportional to and an index of the range of segmentary differentiation in the maximal lineage as it exists now. Eight to ten ascendant generations are usually reckoned between contemporary minimal lineages and the founding ancestor of the maximal lineage of which they are part.

Genealogies are not remembered for their own sake by the Tallensi. They are relevant primarily as the mnemonics of the lineage system, and are bound up with the institutions which demonstrate the formal unity or the internal differentiation of a lineage. Among the social events in which this occurs most conspicuously and frequently are sacrifices to ancestor spirits and mortuary and funeral ceremonies. The latter, especially, epitomize the entire social structure and enable one to see precisely how the lineage system is constituted.

All Tale lineages are hierarchically organized between the limits of the minimal lineage on the one hand, and the maximal lineage on the other. Thus every minimal lineage is a segment of a more inclusive lineage defined by reference to a common grandfather, and this, in turn, is a segment of a still wider lineage defined by reference to a common great-grandfather; and so on, until the limit is reached – the maximal lineage, defined by reference to the remotest agnatic ancestor of the group. A minimal lineage crystallizes out, so to speak, only on the death of its founder. Until then it is submerged in a wider lineage.

Within a lineage of whatever span, each grade of segmentation is functionally significant. Each segment has its focus of unity, and an index of its corporate identity, in the ancestor by reference to whom it is differentiated from other segments of the same order in the hierarchically organized set of lineages. Sacrifices to the shrine of this ancestor require the presence of representatives of every segment of the next lower order; and this rule applies to all corporate action, of a ceremonial or jural kind, by any lineage. This is the fundamental rule of lineage organization.

The solidarity of a lineage is a function of the co-operation of its major segments, the segments of the highest order. In matters that concern the lineage as a whole and in situations which express its corporate unity, members of the lineage co-operate as representatives of its major segments. The lesser segments constituting each major segment receive explicit recognition only in relation to the major segment. They emerge, then, as major segments of a major segment, and their segments, again, emerge only in relation to them, and so on down to the minimal segments. In these affairs of common concern, rights and duties, privileges and obligations, are distributed equally among the major segments of a lineage, thus emphasizing their equality of status.

A lineage system of any span emerges in any of its activities as a system of aliquot parts, not as a mere collection of individuals of common ancestry; it represents an equilibrium maintained by the relations between its constituent parts. Thus its internal differentiation is of a limited and balanced kind. That is why only the minimal differentiation – a division into major segments –

is recognized in the corporate activities of a lineage. In Tale theory, the source of this primary segmentation of the lineage is the minimal lineage, in which a group of brothers (two or more) are *differentiated* from one another as individuals, each of whom is the potential founder of a new lineage, but are *united* through their relationship to their common father. A pair of brothers having the same father (*sunzɔp*, sing. *sunzɔ*) are the potential originators of a pair of major segments of the lineage deriving from their father. So all major segments of a lineage of any span are visualized as being derived from brothers, sons of the lineage founder, and are described as *sunzɔp* to one another. Their lineage ties, summed up in the concept *sunzɔt* (brotherhood), are thought of as being founded on, and derived from, the ties which hold between brothers. By the rule of exogamy, women members marry out of the lineage, so they do not contribute to its perpetuation and do not affect its organization.

The growing point of the lineage is the family. It is always patrilocal; ideally, it is also polygynous. Thus children of the same man are differentiated from one another as children of different mothers. This provides a further criterion of differentiation within the lineage. Like the sons of different mothers and the same father, segments of a lineage may, in certain situations, be distinguished from one another by reference to their different progenitrices. In other situations, like sons of one mother, segments of a lineage may be grouped together by reference to a common progenitrix.

The rule is that whenever a lineage emerges as a corporate unit in its own right and not as a segment of a more inclusive lineage, it is identified by reference to its founding *ancestor*. The prototype of this is the notion of all the children of a man forming a unit of common descent in relation to their father, whether they are the children of one wife or of several wives. But when a lineage emerges as a segment of a more inclusive lineage it is identified by reference to its founding ancestor's *mother*. This serves to differentiate it from co-ordinate segments of the same patrilineal origin in the same way as the children of one man are genealogically distinguished *inter se* by reference to their several mothers. Within the patrilineal joint family the children of one woman have a much stronger bond of solidarity than the children of one man by different mothers. By derivation from this the identification of a lineage by reference to its progenitrix emphasizes its corporate solidarity in contraposition to other segments of the same patrilineal descent; whereas identification of a lineage by reference to its founding ancestor emphasizes its internal segmentation. The following diagram (Fig. 1) illustrates this rule.

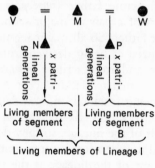

FIG. I

x patrilineal generations of descent may be from one to seven or eight generations. Lineage I is identified as a corporate unit by reference to its founding *ancestor* M. Segments A and B are identified as segments of Lineage I by reference to their respective founding ancestresses, V and W.

A maximal lineage cannot, by definition, have any *sunzɔp* identified strictly in terms of common agnatic descent. If it had it would be a segment of a more inclusive lineage. The ancestor who founded it must be unique, in retrospect. He may be accounted for by a myth, but his unique place in the genealogical tree can sometimes be naturalistically explained. If the formation of recent lineage segments is studied, it can be seen that segments sometimes die out, owing to the extinction of a branch of the lineage. When this happens, the surviving segments reach a new equilibrium. The defunct collateral line is forgotten in due course, together with its genealogy, for this no longer has any structural significance. Thus an existing maximal lineage may be merely the surviving segment of what might have been a wider maximal lineage.

It is characteristic of Tale social organization, however, that maximal lineages do often have *sunzɔp*, either through ties of clanship, or by incorporating, or being joined to, lineages not strictly of the same agnatic descent.

A lineage is a temporal system in equilibrium; continuity in time is its fundamental quality. Its constitution and dimensions at a given time represent a phase of a process which, as the Tallensi see it, has been going on in exactly the same way from the beginning of their social order and is continuing into the future. The contemporary phase of a lineage is more than a product of its past; while embodying all the significant changes that have occurred in it throughout its past, it is, at the same time, the embryo of its future organization. A Tale lineage cannot be dissociated from its temporal extension.

A maximal lineage is fixed with reference to its founding ancestor, who is the focus of its unity and the symbol of its corporate identity. From time to time, its unity and identity become explicit in the common cult of this ancestor and in the regulation of intra-lineage relations which hinges on it. The ancestor cult is the calculus of the lineage system, the mechanism by means of which the progressive internal differentiation of a lineage is ordered and is fitted into the existing structure. It is also the principal ideological bulwark of the lineage organization.

Every maximal lineage is continually expanding and pro-liferating through the fission of its segments. But though its span is thus constantly increasing, its form does not alter. It has a fixed centre and a fixed locus. It always remains the same lineage, and, in theory, new maximal lineages cannot arise through the splitting up of an existing maximal lineage. It is a unit of common agnatic descent and no branch of it can ever repudiate this. In theory, ties of descent can never lapse. A maximal lineage is also an exo-gamous unit, and no branch of it can contract out of this bond.

To the contemporary observer, the Tale lineage system appears as a stable and finally established factor of the constituted social structure, the ground plan of which was laid down in the distant past by processes which are still going on. In any maximal lineage, therefore, the fission of minor segments does not alter the equilibrium of the major segments at any given time. Changes in minimal lineages, in fact, cannot alter the equilibrium of any segment greater than the minimal. This is reflected in the naming of lineages and their segments. Every lineage is named after its founder, and as long as a lineage persists it bears the same name. Thus the names of segments greater than the minimal may be regarded as fixed once for all. A lineage is called the 'children (*biis*)

of so-and-so' or 'the house (*yir*) or people of the house (*yidɛm*)
of so-and-so'. These ways of denoting a lineage are interchange-
able. A lineage may also be described as a 'room' (*dug*) of a
superordinate lineage. This indicates that it is being thought of as a
segment of the latter.

This formal analysis will be more easily understood from the
diagram (Fig. 2) which shows in a simplified way the structure
of a major segment of a maximal lineage which is assumed to
have only two major segments. The one not included must be
homologous with the major segment depicted here, for a lineage
necessarily divides into segments of equal order. The relations
between the two major segments of our hypothetical maximal
lineage are repeated, at a lower order of segmentation, in the
relations between the sub-segments of each major segment. In
the diagram the relations of sub-segments X and Y to each other,
within the framework of major segment I, are identical in form
with those of major segment I and its brother (*sunzɔ*) segment,
major segment II, and this rule holds for every subsequent order
of segmentation. The interests, rights, and obligations associated
with such an ordered series of homologous intersegment relations
vary in kind and number and are graded in precision.

If we describe a major segment of a maximal lineage as a
primary segment, then segments of the order of X and Y in Fig. 2
may be called *secondary* segments, and Y is split into two sub-
segments of the next lower order (*tertiary* segments) V and W.
These have, respectively, two (*g* and *h+i*) and three (*k*, *m* and
n+o+p) segments of still lower order; and some of these consist,
as the diagram shows, of subordinate segments of the lowest
order, i.e. minimal segments. X has no segments co-ordinate
with V and W, but has four segments of the next lower order
(*a+b*, *c*, *d+e*, and *f*) some of which are divided into minimal
segments.

These are all ways of grouping the same thirty-five men in a
set of lineages of common agnatic descent – women members
being excluded from the diagram for the reason previously stated.
In some situations, each grouping emerges as a lineage in its own
right; in others it acts as a segment of a lineage of a higher order.
When, for example, the 'house' (*yir*) of F.VI.1 (i.e. *a+b*) assembles to sacrifice to his spirit, it acts as a self-contained corporate
unit, a lineage identified, in this situation, by reference to its

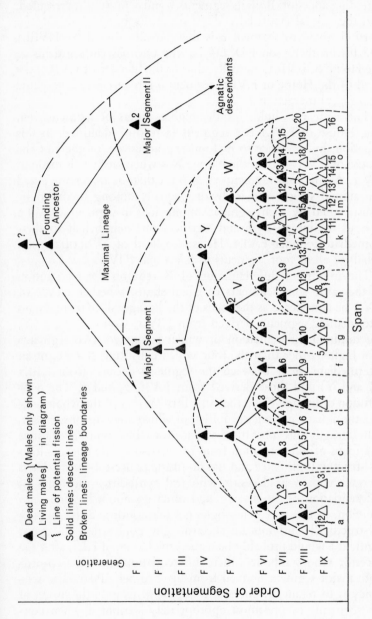

FIG. 2 Diagram illustrating the Paradigm of the Lineage

founding ancestor. Both its segments, *a* and *b*, must be represented, and the animal sacrificed is divided equally between them. The portion given to segment *a* is then equally shared by F.VIII.1 and his brother's son F.IX.2 – i.e. the two nascent segments of the House of F.VII.1. Actually, during the lifetime of F.VII.2, the head of the House of F.VI.1, segment *a* remains only a potential lineage, submerged in *a+b*.

The same rule holds for the higher orders of segmentation. The House of F.VI.1 is a segment (a 'room' identified, in this situation, by reference to its founding ancestor's mother) of the House of F.V.1 – i.e. of the lineage X – when sacrifice is made to F.V.1. All the four segments of X must then be represented, and the animal slaughtered is divided equally among them, to be redistributed by each segment among its sub-segments. F.VII.2 is head of lineage *a+b* and sacrifices on their behalf to their immediate founder F.VI.1. He is also head of X sacrificing on behalf of all its four segments to F.V.1 and F.IV.1.

The minimum differentiation of X requires the recognition of the same four segments of equal status, whether F.IV.1 or F.V.1 be considered as the focus of the lineage, since there are no intermediate segments derived from F.IV.1. Contrast lineage Y, the minimal differentiation of which recognizes two segments of a higher order than the four which constitute X – i.e. of an intermediate order between the segments derived from F.III.1 (X and Y) and those derived from F.V.1, 2, and 3. The constitution of lineage X has the same form as that of the whole lineage, the agnatic posterity of F.III.1, of which it is a major segment.

In lineages thus constituted an ancestor who has no significance as a focus of differentiation from other co-ordinate lineages loses his structural relevance and has no distinct ancestor-shrine. He is merged with a predecessor who still symbolizes the focus of differentiation of the lineage, and often no doubt he fades into oblivion. In this way genealogies get telescoped and the ranking of structurally insignificant ancestors gets confused. Thus major segment I would probably be named the House of F.I.1, after the ancestor who symbolizes their differentiation from the co-ordinate major segment derived from his 'brother', F.I.2. The tendency is to retain the ancestor to whose generation the fission of the segments can be most appropriately attributed when considering their present structural relationship. The naming, here, would fit in with the relations of the two major segments as representing the primary differentiation of the maximal lineage one generation nearer than the founding ancestor. Though F.II.1 and F.III.1 are still invoked in sacrifices, their order of succession tends to be confused. F.II.1 will lapse more easily than F.III.1, who will be remembered as the actual father of F.IV.1 and F.IV.2.

Similarly, lineage X will be described as the House of F.IV.1 when its status as a major segment of the House of F.I.1, co-ordinate with Y, is the main issue. But members of its constituent segments might also describe themselves collectively as the House of F.V.1, after the most recent ancestor from whom their lines of descent diverge, when stressing their connection as 'rooms' of a single lineage. Which nomenclature is used depends on the situation and on the range of the lineage concerned in it.

These usages illustrate a deep-seated trait of Tale thought, the naming of social units by contraposition. They explain, also, why no historical validity can be attached to Tale genealogies, beyond the time of the great-grandfathers. A genealogy maps out a particular set of lineage relations, it is not a true record. Thus it is impossible to ascertain, as it is from the Tale point of view irrelevant to ask, whether or not any ancestors intervened between the founder of a maximal lineage such as is here sketched and the originators of its existing major segments. If there were any, it is quite possible that they have been dropped from the genealogy because they were redundant, in terms of the structure of the lineage. Similarly, the ancestor whose name is preserved as the founder of the maximal lineage may have been preceded by others whose names have lapsed from tradition for the same reason.

Lineages of the same order of segmentation are not all of equal span. Thus, lineage c – the House of F.VI.2 – is co-ordinate with lineage $a+b$ – the House of F.VI.1 – but of a lesser span. Lineage X, similarly, has a lesser span than lineage Y. The span of a lineage is a measure of its internal differentiation, whereas its order of segmentation defines its relations, as a corporate unit, with other units of a like sort.

The distinction is important in the conduct of lineage and community affairs. Co-ordinate segments are *sunzɔp* to one another, irrespective of span; but a lineage of small span is more closely integrated than one of large span. Thus the major segments of our diagrammatic maximal lineage are *sunzɔp*, being descended, as it is thought, from a pair of brothers having the same father; so are lineages X and Y; so are V and W; so are the four constituent segments of X, the two constituent segments of V, and the three constituent segments of W; the same rule holds for the lesser lineages down to a pair of brothers, such as lineage *c*, who are also *sunzɔp*.

It is evident that *sunzɔp* can be graded according to their genealogical distance from one another. The Tallensi say, *ti a ba yɛnni biis* (we are the children of one father) when they want to stress common ancestry, or *ti a yaab yɛnni biis* (we are the children of one ancestor) whenever they emphasize remote common ancestry. The same distinction is expressed in the appellations *dug* (room) and *yir* (house). Members of lineage V, for example, will speak of themselves as *dugdɛm* – members of one room – by contrast with members of lineage W, whom they would describe as *ti yidɛm* – people of our house. Members of X will say, 'we, the children of so and so (F.IV.1) are four rooms', but they will speak of themselves as *dugdɛm* in contrast to Y, who would be called their *yidɛm*, for example, at funeral ceremonies. The designation *'dug'* is commonly associated with the identification of the lineage by reference to its progenitrix; the term *'yir'* with its identification by reference to its founding ancestor. Thus the segments of X or of Y are closer *sunzɔp* than are X and Y, which, again, are closer *sunzɔp* than are the major segments of the maximal lineage. A specially close tie is acknowledged also between lineages of common agnatic ancestry which have a common ancestress. Thus, if F.V.1 had three wives from whom have sprung respectively, segments *a+b* and *c*, *d+e*, and *f*, then the first two segments regard each other as closer *sunzɔp* than either of the other two.

The connection between genealogical distance and the *sunzɔt* tie is most clearly seen in situations which bring out the structural equilibrium of the lineages. When, for instance, a funeral occurs in segment *g*, contributions of cooked food will be made by the *sunzɔp* as follows: by the head of *h+i*, the closest *sunzɔ* segment, on behalf of that segment; by the head of W on behalf of the whole lineage, not by its constituent segments, the contribution being given to its *sunzɔ* V, not to the segment *g*; and by the head of X, if it is the funeral of an elder, on behalf of the whole of X, when the recipient unit is considered to be Y, not *g*. If the other half of the maximal lineage, major segment II, sends a contribution, it will be given by the head of the segment on behalf of the whole unit as an obligation owed to the *sunzɔ* segment, the House of F.I.I. Cooked food, beer, and the meat of ceremonially slaughtered animals are distributed according to the same scale of genealogical distance, and the allocation of ritual duties, though less strictly regulated, depends on it, too. Hence any members of W who may be present at the funeral are regarded as representing W, however many segments of W they may individually represent. They are entitled only to those portions of food and beer and meat which are the due share of W. An identical share may be claimed by members of X who are present, whether they come from one of its segments or from all; and the same holds for the major segments of the maximal lineage. *Sunzɔt* here implies duties and privileges vested in the lineage as a corporate unit and exercised on its behalf by any representative of that unit in relation to other units of a like sort.

The tie of *sunzɔt* operates in the same way and is recognized according to the same rules in all corporate actions and in relation to all corporate interests of a lineage. It dictates the manner in which every activity associating people in groups is organized. Hoeing and building teams organize their work roughly along the lines of the lineage structure, and distribute the food they receive as reward in accordance with it. In a hunting party, lineage *sunzɔp* have a right to 'pull out' (*fɔ*) a foreleg of any animal killed by a member of a *sunzɔ* lineage. But here we have been concerned only with the paradigm of the lineage system, a mere definition of the fundamental concept of Tale social organization.

There is one last point which should be stressed. When a lineage emerges in corporate activities it functions, not as a collection of individuals, but as an internally differentiated structural unit. A man does not take part in such activities as a member of the whole agnatic descent group involved, but as a member of a particular segment of the lineage concerned. The components of a lineage are visualized as lineage segments, not as individuals, even though, in a particular situation, some or all of the lineage segments may be represented by only one person each. Thus to turn to the diagram (Fig. 2) the thirteen male members of lineage segment X are not all on equal terms in corporate activities involving this lineage as a whole. The individual designated F. VIII.9 himself constitutes a sub-segment (*f*) of this lineage, of the same order as *a* and *b* combined. He has rights and duties corresponding to those vested in the whole sub-segment *a*+*b*, and is therefore on an equal footing with the head of that sub-segment, F.VII.2. The individuals F.IX.1, 2, and 3 have no personal *locus standi* in corporate activities; they take part as members and representatives of the sub-segment *a*+*b*.

THE CONNECTION BETWEEN KINSHIP AND THE LINEAGE SYSTEM

We have described the morphology of the lineage system without reference to the domestic organization of the Tallensi, but these two planes of social structure cannot be isolated from each other in the actual life of a Tale community or individual. The Taləŋ lives his life as a member of a lineage and a member of a family. It is true that the interests and ends – primarily economic and reproductive – which the family subserves differ significantly from those – primarily jural and ritual – which the lineage subserves. But the two categories are co-ordinated and integrated. The individual's rights and duties, sentiments and values, manners and moral conduct, all his thinking, feeling, and acting in the context both of lineage and of family relations, are organically integrated, as are lineage and family.

The interpenetration of lineage and family appears most clearly from the way segmentation in the domestic family is parallel to and regulated by segmentation in the lineage. Families divide along the lines of lineage cleavage. Their local and functional grouping corresponds to the agnatic distance from one another of the family heads, and their social relations with one another depend upon this fact.

Since kinship forms the nexus between the domestic organization of the Tallensi and their lineage system, they apply the concepts of kinship to describe and define both domestic and lineage relations. In the social structure as a whole, kinship is the fundamental bond. It furnishes the primary axioms of all categories of inter-personal and inter-group relations.

From the standpoint of the individual, all the norms and conditions which govern his social behaviour fall within a single, syncretic frame of reference, of which kinship forms the base-line.

The alignment of individuals and groups in accordance with material, jural, or ideological interests follows genealogical lines. The Tallensi do not have one category of social relations for economic ends, another for jural and political purposes, and a third for religious purposes, with only adventitious connections between the three. Their economic, jural, political, and religious institutions interlock and determine one another; and they do so because they have a common foundation in the genealogical structure of the society. With few exceptions, social relations among the Tallensi always have a genealogical coefficient.

It is essential to distinguish two kinds of genealogical ties among the Tallensi. There are, firstly, cognatic ties, ties of actual or assumed physical consanguinity and of the social relations entailed by them, which link person to person or an individual to a lineage or one lineage to another in a specifically defined and particular bond. Such ties may be traced through males only, through females only, or through both males and females. Secondly, there are lineage ties, social relations based on the tie of common agnatic descent.

Cognatic kinship and agnatic kinship have different, and in some situations, even opposed functions in Tale collective life. The distinction turns on the principle that agnatic ties unite people in corporate groups serving common interests and held together by common values, whereas cognates do not necessarily[1] form corporate groups. Cognates have mutual interests, bonds of sentiment and of reciprocal obligations, but not necessarily common interests. All the members of a given lineage have the same agnatic kin and therefore identical lineage ties; but only identical siblings – that is, brothers or sisters by the same parents – have the same cognatic kin, and then only until the time when they become parents. By making him a member, automatically, of a maximal lineage and clan, his agnatic descent fits the individual into the constituted framework of Tale society. This gives him also a special field of defined social relations with clear contours. Cognatic kinship creates a number of contingent social ties for the individual. Unlike lineage ties, they differ in quality; they change in the course of his lifetime; in theory, their range

[1] It must be remembered that agnates are also cognates; cognates *who are not agnates* do not form corporate groups.

is indeterminate, since there is no limit to the reckoning of cognatic kinship. Tallensi often discover kindred of whom they were previously ignorant in the most unexpected places in their own country, even in their own clan. But it is extremely rare for anyone to discover a hitherto unknown clansman even in these days of relatively great mobility.

Cognatic kinship has an extraordinarily wide ramification among the Tallensi. They maintain, and genealogies of individuals bear this out, that if enough were known of the genealogical relationships of the people of adjacent settlements, they would all be found to be related to one another. In the clan itself, under the surface of its strict patrilineal organization, the filaments of cognatic kinship bind individuals together by special personal bonds which operate independently of lineage ties. This complex and unlimited ramification of cognatic kinship gives these relations great fluidity, in contrast to the comparative fixity of lineage relations. Individuals are often connected by multiple cognatic ties which are differentially effective in different situations; and cognatic ties make breaches in the barriers of lineage and clan exclusiveness, thus extending widely the flow of social relations.

The domestic family is the matrix of all the genealogical ties of the individual, the contemporary mechanism for ever spinning new threads of kinship, and the focal field of social relations based on consanguinity. In it we can observe the working of the nuclear patterns of kinship and the formation of the ideas and values which steer the individual in all his genealogical relationships. In the domestic family we get the sharpest picture of the interaction between cognatic kinship and agnatic ties. We have there the elementary ties of cognatic kinship linking parent to child and sibling to sibling, and we have also the agnatic tie which sets apart the males as the nuclear lineage. We can see the centrifugal pull of matrilateral kinship counterbalancing the centripetal pull of patriliny, and so producing the primary equilibrium of Tale social life.

Through his primary relations of consanguinity in the domestic family, the individual is linked to agnates of his clan in other families and to cognates in other families and other clans. There is, however, one significant difference between intra-familial and extra-familial bonds of kinship. Unlike the latter, the former are double-sided, each relationship containing within itself both an agnatic and a cognatic component. In a joint family of three generations, for example, a child's relationship with its father differs from its relationship with its grandfather or its mother. These are all cognatic relationships. There is a tension and latent rivalry in the relationship of father and son which stands in marked contrast to the comradeship of grandfather and grandson. On the other hand, grandfather, father, and son belong to the same lineage, and are united by a common interest in patrimonial land and in the product of their joint labour, by a common ancestor cult, by rights and duties vested in that lineage in relation to like segments of a greater lineage, and by their common concern for the continuity of their line. These are part of the whole complex of common interests and values which mobilizes corporate action and maintains corporate solidarity in the maximal lineage and clan of which they form a minor segment. Men of a joint family may have divided loyalties with reference to their cognatic relationships, but they act in union and on behalf of one another and of the whole unit in virtue of their lineage ties.

There is another paradox in the structure of the Tale family. Consanguineous relationships arise out of parenthood; but parenthood presupposes marriage, a union of a man and a woman who are, by definition, not kin.[1] The bonds of kinship are rooted in a bond the very essence of which is the absence of kinship. Hence comes the ambivalence inherent in intra-familial cognatic relations. Hence comes, also, another category of social relations both in the family and between members of different families and clans: relations of affinity. There are two axes, as it were, in the structure of the family, the axis of kinship and that of marriage.

The constellation of ties and cleavages which makes up the focal field of kinship is conditioned by the formative principles and the values which shape the greater society into which the domestic family fits and which it helps to knit together. All person-to-person relations in the family are biased by the values attached to patrilineal descent and to maternal origin in the total social structure.

[1] This point is also stressed by Forde (1941).

THE GENERIC CONCEPT OF KINSHIP

The generic concept of kinship among the Tallensi, expressed by
the word *dɔyam*, subsumes all kinds and degrees of genealogical
relationship, however remote, through one or more progenitors
or progenitrices. But its primary reference is to procreation. One
might perhaps translate it by the word 'generation' in its etymo-
logical sense. Its root, the verb *dɔy*, to bear or beget a child,
signifies both the male and the female function in procreation.
Dɔyam, the abstract noun, describes the process, or the act, or
the biological capacity, of bringing a child into the world, as
well as the ties thus created. The logic behind this concept of
kinship is plain. Every genealogical relationship goes back eventually to one pair of parents. Both in fact and in Tale kinship
theory, the parent-child bond is the nodal bond of kinship.

Now though the Tallensi see every genealogical relationship
as a tie of physical consanguinity, this is not what matters most.
More important are the social relations entailed by consanguinity.
A genealogical tie between two people, or two genealogically
defined units, comes into action in the palpable facts of economic
life, jural relations, moral values, ceremonial duties, and ritual
ideas and performances. Genealogical ties very often have economic or political utility and are sustained by powerful moral
and religious sanctions. Much of Tale kinship custom is specific.
In ceremonial situations, different categories of kin often behave
in prescribed ways which sharply distinguish them from one
another. But in the routine of ordinary life the intercourse of
kinsfolk has no such formal pattern, and is often indistinguishable
from the intercourse of friends, neighbours, or other associates.
The significant thing, for the Tallensi, is not so much the prescribed pattern of conduct as the fact that all kinship is morally
binding.

The Tallensi indicate this most clearly in their attitude to
marriage. They draw a sharp distinction between *dɔyam* (kinship)
and *dien* (in-law-ship). The crux of the rule of clan exogamy and
of the collateral prohibition of marriage with any person con-
sanguineously related – and no distinction is made between these
two classes of restrictions – is the principle that *dɔyam* and *dien*
are irreconcilably contradictory.[1] Marriage implies the absence

---

[1] *Cf.* my paper (1936) previously cited.

of kinship ties between the parties, and kinship the impossibility of marriage. Kinship ties exist in their own right; *dien* is an *ad hoc* alliance of a contractual nature. The parties to a marriage deliberately and voluntarily enter into a bond, and are bound by rights and duties which did not exist before. Kinship ties are not voluntary, and are automatically binding, entailing moral, ritual, and jural obligations. In the relations of in-laws there is an avowed element of mutual coercion; they rely upon a special, impartial jural instrument, the bride-price, for the adjustment of their rival claims on the woman.

Tallensi think of *dien* as a relationship forever fraught with the possibility of conflict over these claims. Nearly all their litigation nowadays, like much inter-clan fighting in the past, concerns bride-price debts or the rival claims of father-in-law and son-in-law over a woman and her children. Such quarrels are incompatible with kinship; they would cut to pieces the solidarity of kinsfolk. That is why, say the Tallensi, kinsfolk are prohibited from marrying.

Dɔyam, in short, presupposes some degree of identification of the parties concerned, mutual or common interests, and especially a bond of amity excluding strife which might fix a permanent gulf between them. These norms of kinship the Tallensi regard as axiomatic, as the *a priori* moral premisses of their social behaviour. Behind the utility of kinship in practical life, and the jural and ritual sanctions that buttress kinship ties, stands the notion of kinship as the rock-bottom category of social relations, inviolable in its own right. The root of this notion lies in the theory of the parent-child bond, and it reflects the fact that genealogical relationship is the binding medium of Tale social structure. Tallensi take it for granted that the ideal norms of kinship will often be violated, but the lapses of individuals and the fact that strong sanctions exist to check them, do not diminish the absolute, *a priori* character of kinship.

It is worth noting that economic reciprocity – which is a conspicuous element of kinship, especially among close kin – and jural rights and duties have a much less direct connection with this fundamental assumption of kinship than have religious bonds. This is connected with the important place of ancestor worship in Tale society. In order to trace genealogical relationship it is necessary to recollect common ancestors; and if these have

religious value, ritual allegiance to a common ancestor or ancestress inevitably forms an intrinsic feature of kinship ties and
their most powerful sanction. The more distant a genealogical
tie is, the more does it become a matter of moral and ritual, and
not of jural or economic, relationship.

The idiom of kinship has such a dominant place in Tale thought
that all social relations implying mutual or common interests
tend to be assimilated to those of kinship. This happens, for
example, with ties of local contiguity and of politico-ritual
interdependence.

The starting-point for the wide recognition of kinship among
the Tallensi lies in their notion that conception is impossible
without sexual relations. In their view man and woman have an
equally vital role in the act of conception. This is summed up
in the maxim, 'A man and a woman together procreate a child'
(*Bumpɔk ni buraa n-kab dɔya bii*). Hence kinship through one's
mother and through any of her ascendants, counts equally with
kinship through one's father and one's paternal line, though in
different ways.

THE SIGNIFICANCE OF PATERNITY AND PATRILINY

A person's genealogical ties are fixed by his parentage. The rights
and duties which are critical for his role and status in society all
stem, in the last resort, from the fact of birth. He cannot divest
himself, or be divested, of the bonds created by his birth and yet
remain a member of the society; nor can he fully and uncon-
ditionally acquire those bonds except by birth. No other ties can
wholly supersede them in linking person to person and affiliating
individuals to defined social groups.

But though a person acquires significant social ties through both
parents, his agnatic relationships have an outstanding importance,
particularly for a man. For not only is patrilineal descent the
vertebral principle of Tale social organization and the vehicle
of the continuity and stability of the social structure, but men
hold the reins of authority, direct economic life, control the
political organization, and are supreme in religious and ceremonial
thought and action. From his father a man derives his right to
inherit land and other property, his clan membership and the
political rights and ritual obligations that go with it, and his

ritual relations with his ancestors. A woman does not inherit or transmit land or other property of value, nor does she succeed to or transmit political or ritual office. But clan membership and the concomitant totemic observances, as well as her ritual allegiance to her ancestors, mean a great deal for her social destiny and for her children. Like many other pre-literate societies, the Tallensi distinguish between physiological and jural paternity – between *genitor* and *pater*, as Radcliffe-Brown puts it. However, the Tallensi feel very strongly that a person's physiological father is his right jural father. The assumption is that all the social attributes which come to a child from his father should come from a father who both begot him and recognizes him as legitimate offspring. Thus there is a tendency for conflict to arise if a person's physiological and jural paternity do not coincide. Such a person tends to be penalized in respect of his social status, for jural paternity by itself cannot altogether take the place of physiological-cum-jural paternity.

The conflict is least severe in the case of an adulterine child, who is always accepted as the rightful child of its mother's husband. No difficulty at all arises in the case of a daughter, since she marries out of the lineage as soon as she is nubile. An adulterine son has complete and unreserved filial status. He has full rights of inheritance, of succession, and of ritual access to his putative patrilineal ancestors. Nevertheless, the blot on the scutcheon is not without effect; it is often the underlying cause of friction between a man and his jural, but not physiological, father or brothers.

This kind of friction is said not to occur in the infrequent case of a child begotten in permitted extra-marital intercourse, which the Tallensi do not regard as adultery (*poɣambɔn*). A man who is sterile or impotent may allow his wife to conceive by another man. Tale sentiment takes no cognizance of physical paternity in this situation; one might almost describe it as, from their point of view, a kind of artificial insemination.

But take the case of an unmarried girl's child (*ɣi ɣiem bii*) which is brought up as the foster-child of her father or brother. If the child is a girl, no difficulties will occur, as she will eventually get married and Tale kinship ideas permit her children to regard her mother's lineage as their nearest matrilineal kin. She will even be permitted to marry into her mother's clan, provided it is a different major segment of the clan. Many instances of this are on record, for no stigma attaches to anybody concerned.

An illegitimate boy will grow up to all intents and purposes as a child of his mother's father's or brother's family, and as a full member of the clan. He will be treated exactly like other sons of the family, whom he will call his brothers. But he has two fundamental disabilities which nothing can overcome and which pass, by the principle of the corporate identity of the lineage, to his agnatic male descendants. He has no right to inherit his quasi-father's patrimonial estate, though he may be permitted to do so failing other heirs; and he has no right to sacrifice directly to the latter's spirit, or consequently, to the lineage ancestors. He cannot, therefore, succeed to the custody of the lineage ancestor shrines or hold an office vested in the lineage. The tie of true agnatic descent, which alone confers these rights, cannot be fabricated. An illegitimate boy's status in the lineage is, in fact, therefore, inferior to that of his quasi-brothers. He is always liable to a certain amount of covert contempt, though good manners forbid mention of his irregular descent. His lot may be harder, if he has an intractable character, than that of a sister's son (*ahaŋ*) living with his mother's brother (*ahɔb*), for the latter has a home of his own to fall back upon, where he is entitled, by right of legitimate descent, to his patrimonial inheritance, protection, and communion in the ancestor cult. In no circumstances, however, can an illegitimate son be deprived of those rights that belong to him. His quasi-father's home is his home, his mother's clan his clan. From him may spring an accessory lineage of the clan. In theory, also, he may marry a daughter of his adoptive clan, but I know of no case in which this has occurred; presumably public sentiment would be against it.

The status of a male slave (*da'abɔr*) in former times was similar, though in some respects less advantageous. In order that slaves might be fitted into the social structure, they were incorporated into the family and lineage of their owners by being placed under the spiritual guardianship of one of their owner's lineage ancestors' shrines. This gave them a quasi-filial status of the same kind, with the same rights and disabilities, as that of an illegitimate son. Analogously, the status of a slave girl (*da'abpɔk*) resembled that of an illegitimate daughter.

Tallensi declare that it would be disgraceful and immoral to treat a slave or an illegitimate son or brother differently from true agnatic kin. But they also affirm that, nevertheless, a slave or an illegitimate son cannot possibly become a son or a brother in the full sense of those terms. He can be to you *in loco filii* or *fratris*, but not *filius* or *frater*, since he cannot sacrifice directly to his quasi-father's spirit, or inherit, as of right, his patrimony. The principle at issue is that physical father and social father, *genitor* and *pater*, must be the same person. The vicissitudes of marriage illustrate the same point. It often happens that a young woman becomes the bride of one man, is later married to another man, by whom she has a child, and finally returns to settle down with her first husband. The second husband (or his heir) then has the right, at any time during its life, to redeem his begotten child by paying over to the woman's father a proportion of the bride-price equal to one cow, whether or not he has completed the formalities of marriage. His right rests upon the fact of the child's acknowledged paternity. Moreover, a man who was begotten by a former husband of his mother and grows up in the home of her final husband but is not redeemed by his *genitor*, has the right to go to his *genitor*'s home and claim all the rights of a son. To regularize the position, he may himself pay the cow of redemption to his mother's father's heirs.

THE RITUAL COEFFICIENT OF PATRILINY

All these cases of irregular paternity bring out a point of central importance in Tale thought. The bonds of physical consanguinity and the concomitant social ties created by parenthood are *ipso facto* spiritual and moral bonds. They have a ritual coefficient in which the people see their deepest meaning. One's bond with one's patrilineal ancestors is a ritual bond with them and with the other mystical forces associated with the existence and well-being of the lineage. On them depends the course of one's whole life. Legitimate paternity is a *sine qua non* for a right relationship with one's agnatic ancestor spirits. There is a spiritual hiatus between an illegitimate son or a slave's son and his adoptive lineage ancestors. Being only accidentally, as it were, under their tutelage, he cannot use them to master the vicissitudes and frustrations of life as freely as may a true scion of their line. Political and ritual office can be held only by men who are legitimate members of the lineage in which these offices are vested; and the sanction for this lies in the key functions of these offices in the cult of the ancestors and of the other mystical powers coupled with the lineage.

All social relations in which agnatic descent has a decisive function have this ritual coefficient. Legitimate paternity is bound up with the exclusive rights over his wife's reproductive powers conferred on a man by marriage. A child's legitimacy is jeopardized if it is conceived or born in a situation in which its father's marital rights do not prevail. Hence, sexual intercourse between a man and his wife (and, by extension, between any man who is a near agnate of the husband and any woman who is a close agnate of the wife) at her paternal home is rigorously tabooed. For in her father's house a woman has the status of a daughter, and comes under her father's authority and the tutelage of his ancestors. For the same reason, a child must not be born in the house of its maternal uncle. Its ties with its maternal kin must not be confounded with its bonds with its paternal kin, for these represent opposed forces in its social world, and matrilineal kinship ties must not be allowed to become a threat to lineage ties. It is to obviate this that custom forbids a man to demand his daughter's bride-price if she is pregnant.

A woman bears a child to her husband, an additional member for his lineage and clan. When she conceives, it is the right and duty of her husband and his family to care for her; his ancestor spirits must guard her and the unborn child, and the child must be born under its father's roof, literally into the lineage to which it comes as a profit (*yoor*) and a responsibility, and under the auspices of its patrilineal ancestors, the spiritual arbiters of its destiny, to whom it will later in life bring sacrifices. Ideally, it should also grow up under its father's roof.

THE SIGNIFICANCE OF MATERNAL PARENTAGE AND THE UTERINE LINE

Though paternity is overwhelmingly dominant in the jural, economic, and ritual constitution of Tale society, it would be surprising if maternal origin had no institutional function in the social organization and ideology. In fact, as we have seen in our analysis of the lineage system, the two principles of patrilineal descent and maternal origin are always complementary in their action. This is characteristic of their relationship in all Tale institutions.

One indication of the sociological significance of maternal origin is the quality of the bond between mother's brother (*ahɔb*) and sister's son (*ahɔŋ*), either of the first degree or in a classificatory sense. A person's own maternal uncle (*ahɔb*) and maternal grandfather (*yaab*) and their immediate lineage segment impinge directly and frequently on his life. His maternal uncle's home is his own second home; he has a quasi-filial status there, and, in addition to close bonds of sentiment with his uncle and his close kin, he has specific ceremonial rights and duties in relation to them. Though he has no property, succession or inheritance rights in his uncle's home, a man has special material privileges there, which express his quasi-filial status. They show that he is equated with a son in sentiment, while strictly differentiated from a son in jural terms. The kinship tie between sororal nephew and maternal uncle is an important breach in the genealogical wall enclosing the agnatic lineage; it is one of the main gateways to social relations with members of other clans.

But the recognition accorded to the mother's agnatic kin does not bring out the deeper implications of maternal origin as the complement of paternal origin. These implications are contained in the concept *soog* (pl. *saarɔt* or *sooret*) which involves the idea of matrilineal as the counterpoise of patrilineal descent. In its range of efficacy, its mode of operation, its jural and ritual value, and its emotional connotation, matrilineal descent stands in contrast to patriliny.

For obvious reasons, the bond of consanguinity with one's mother is unique and unalterable. It has also a peculiarly personal quality which is emphasized by the fact that one's jural status is not derived from one's mother. Your mother may desert your father and leave you to be reared (*ugh*) by another woman; you may feel some resentment over this all your life, but she remains your true mother, 'the mother who bore you' (*ma dɔyarug*). Nothing can alter this fact or extinguish the kinship ties which arise from it.

The concept *soog* has definite emotional implications, but here we are concerned with its biological and social implications. To the Tallensi, *soog* means a bond of consanguinity which excludes completely any reckoning with lineage affiliation. People who are *saarət* define themselves as 'the offspring of one woman', or, in the actual Tale phraseology, 'to have come from one vagina' (*yi pɛn yɛni*).

Saarət, therefore, are uterine kin, and the uterine tie between mother and child corresponds to the agnatic bond between father and child; children of one woman are *saarət* irrespective of their paternity. A woman may have children by different men of the same clan or of different clans; they are *saarət* irrespective of their clan affiliation.

Soog is more than the bond of personal kinship between mother and child, and between siblings by the same mother. It is the bond of uterine descent perpetuated through the female line in the same way as the bond of agnatic descent is transmitted through the male line. Uterine siblings are *saarət*; uterine sisters' children are *saarət* (but not uterine brothers' children, for the children of men are not their father's *saarət*); uterine female cousins' children are *saarət*; and so on, theoretically without limit, as the following diagram shows:

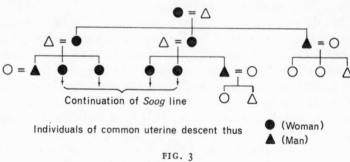

Continuation of *Soog* line

Individuals of common uterine descent thus ● (Woman)
▲ (Man)

FIG. 3

In short, people of common matrilineal descent are socially identified with one another.

As a result of the social recognition of the matrilineal line, *saarət* have defined ways of behaving towards one another, and further, the fact of matrilineal descent is given an institutional embodiment. Matrilineal descent confers distinctive social attributes on people. In this, Tale culture follows the rule found in most other West African cultures where both lines of descent are recognized. The dominant line of descent confers the overtly significant attributes of social personality – jural status, rights of

inheritance and succession to property and office, political allegiance, ritual privileges and obligations; and the submerged line
confers certain spiritual characteristics. Among the Tallensi, it
is easy to see that this is a reflex of the fact that the bond of uterine
descent is an inter-personal bond. It does not subserve common
interests of a material, jural, or ritual kind; it unites individuals by
ties of mutual interest and concern not unlike those that prevail
between close collateral kin in our culture. While it constitutes
one of the factors that counterbalance the exclusiveness of the
agnatic line, it does not create corporate groups competing with
the agnatic lineage and clan. Carrying only a spiritual attribute,
the uterine tie cannot undermine the jural and politico-religious
solidarity of the patrilineal lineage.[1]

The critical spiritual attribute carried by the *soog* line is the
potentiality of being a witch (*sɔi*, pl. *soob* – a term evidently
cognate with *soog*). This potentiality adheres only to the female
line and defines that line in social terms. 'This is my *soog*', a
person usually says when introducing a uterine relative. 'If he
(or she) flies magically (*ayɔt*) I, too, fly magically; if he sees
magically (*nye*) I, too, see magically.' The potentiality of being a
witch is hereditary in the female line. A woman's son and daughter
are both witches if she is a witch. If a person is accused of being a
witch, all his or her *saarɔt* are *ipso facto* accused of this inborn
mystical vice. His agnatic kin, however, are not implicated, not
even his own father or his half-siblings by the same father. These
are the points Tallensi invariably stress, both among themselves
and in answer to an inquirer, if they are asked to define the
meaning of *soog*.

Curiously enough, witchcraft occupies a very minor place in
Tale thought or ritual action, and that is perhaps why Tallensi
make no secret of its hereditary transmission in the female line.
They have no clear notion of witchcraft, no detailed theories
of its mode of operation, and no institutionalized means of com-
bating or sterilizing it. A witch is supposed to fly about (*u
ayɔrɔme*) at night in the guise of a ball of fire; she – for the Tallensi
tend to think of a witch as a woman though they say that men
may be witches – is supposed to be clairvoyant, to see hidden

---

[1] *Cf.* the Ashanti, among whom it is the patrilineal line which has the submerged function (Rattray 1923). Among the Yakö the balance is more even, *cf.*
Forde, *loc. cit.*, 1938.

things; she can recognize fellow witches, foresee death coming to a particular person, prophesy an epidemic, and so forth; she may be beneficent, able to impart luck to a person in a mysterious way, or maleficent, killing people by eating (ɔb) their souls (sii). This is the sum of Tale ideas about the nature of witches.

Witchcraft is something on the fringe of Tale mystical thought and ritual values. I have more than once heard elders of the most conservative way of thinking, to whom the Tale scheme of ritual values is the most important thing in life, declare roundly that there is no such thing as a witch. Though the Tallensi do not put it into words, they obviously feel that the idea of witchcraft is a mixture of superstition and folk-lore, and not a part of the system of ritual values that really matters for the conduct of life. Of course superstition and folk-lore are not negligible. Circumstances occur, now and then, in which people fall back on the idea of a witch's occult aggression to cope with anxieties and frustrations which are normally neutralized through the agency of the ancestor cult; and then a particular person may be accused of being a witch. Significantly, in every case which has come to my notice, some dating back twenty or thirty years, and all of them together numbering less than ten, the accused has been a woman. Most accusations of this sort are made by co-wives, and are symptomatic of the tensions inherent in the joint family.[1]

In the past, the uterine kin of a proven witch, and particularly her children, felt the shame for a time, but they were not publicly stigmatized or penalized. Even the husband of an apparently guilty woman would, judging by what happens nowadays, very often continue privately to believe in her innocence, and even if he admitted her guilt, would not allow this to influence his relations with his children by her.

One reason why witchcraft is of minor significance in Tale mystical thought is because the elaborate ancestor cult and the Earth cult deal adequately with most social and psychological tensions. Another reason is that the idea of witchcraft is not easily reconcilable with the structure of Tale society. In a society so largely dominated by the principle of genealogical relationship, where the residential group is based on the corporate agnatic lineage and where a person's social contacts are almost entirely

[1] This, as is well known, is the rule in very many African societies.

regulated by kinship ties, the norms of kinship necessarily have a very strong hold; and the notion of maleficent secret malice involved in the idea of witchcraft is incompatible with the norms of kinship. It would poison the relationships of kinsmen and throw social life into confusion. As it is, the pegging of witch-craft to the uterine line serves to circumvent any danger to the community. The individuals upon whom an accusation of witch-craft might be expected to throw suspicion are scattered through various lineages and clans. They may protest, but they have no means of taking organized action in defence of their reputation. In any case, an accusation of witchcraft has a very limited range of reaction. Other lineages and families than the one injured by the alleged witch do not feel the injury and even if they harbour uterine kin of the accused person, self-interest prompts them to disregard the implications of the accusation. The folk-lore of witchcraft proves loopholes for such adaptations of logic to social realities. Some say, for example, that a person only becomes a bad witch if his or her mother gives him or her 'witch's medicine' to eat in infancy.

The hereditary transmission of the capacity of being a witch is not connected with the blood (*ziem*), according to the embryo-logical ideas of the Tallensi. Both parents transmit their blood to their offspring, as may be seen from the fact that children may resemble either parent in looks. It is solely a matter of the eye (*nif*, pl. *nini*) – not the physical eye, but the inner, clairvoyant eye which witches are supposed to possess.[1]

As we have observed, the fundamental distinction between uterine kinship and agnatic descent is that the former does not give rise to corporate units of social structure. *Soog* is essentially a personal bond uniting individual to individual. The model of the relations between *saarɔt* is the affection and the feeling of mutual identification that prevail between mother and child, or between full siblings of the same sex. Its keynote is mutual trust as between equals. It is never thought of as a matter of rights

[1] The social definition of uterine descent in terms of the notion of witchcraft is consistent with its functions in the structure of Tale society. The meaning of *soog* is neatly symbolized in this notion. And this is true, also, of the only other attribute of social personality that is held to follow the uterine connection. This is summed up in the concept of *tyuk*, indirect or inadvertent guilt for another person's death or misfortune. It is unnecessary to discuss this here, as it follows the same rules as the idea of witchcraft.

and duties, but as a matter of sentiment, of spontaneous goodwill and intimate, personal confidence. Hence the term *soog* is often metaphorically used to refer to someone with whom one has a relationship of a similar sort, or to denote a very close relative of a class of relatives, whether or not they are true *saarǝt*.

This personal quality in the relationship of uterine kin is brought out in the attitude of the Tallensi to sexual relations between *saarǝt*. Such relations are incestuous if the couple know they are *saarǝt*, however distant their uterine connection. It is like incest between brother and sister and is viewed in the same way, with disgust rather than horror. Sexual desire for a known *soog* relative is so utterly inconsistent with the general attitude of uterine kin towards each other that it is unthinkable in a decent person. It is a taboo (*kihǝr*) which normal people take for granted. There are no ritual or jural sanctions against incest with a *soog*, and Tallensi would say that there is no need for them. The bonds of uterine kin are not subject to overt sanctions. 'I will not refuse him, I will not deny him anything', says the Talǝŋ of a *soog*, thinking of this as an axiomatic characteristic of *soog* kinship.

A person may have *saarǝt* among his own clansfolk, but the majority of his uterine kin are dispersed among other clans than his own. This is a natural result of the rule of clan exogamy. But Tallensi tend to marry into nearby clans rather than into distant ones. A woman of clan A marries into the adjacent clan B; her daughters marry into clans C and D; their daughters into clans E and F; and so the process goes on. The result is, firstly, that with the passage of the generations the uterine line becomes more and more dispersed; and, secondly, that any person's genealogically nearest uterine kin are also spatially nearest to him. In practice, it is these near *saarǝt* that he knows and knows of; it is they who are concerned in witchcraft accusations and *tyuk* guilt. His more distant *saarǝt* – both in the geneaological sense and in residence – are often unknown to him. The main reason for this is that uterine kinship remains effective only through personal contact. The children of sisters know one another because their mothers have kept in touch, visited one another, and helped one another all their lives. Similarly, the uterine grandchildren of sisters know one another, if they live near one another, because their mothers have kept in touch, though not so faithfully as true siblings. Often these contacts are maintained into the fourth generation of a woman's uterine descendants, but rarely beyond it; and a person does not, as a rule, know all the *saarət* whose link with him or her is a common great-grandmother. Some of them will, as likely as not, belong to neighbouring tribes, and he will not even know of their existence. In principle, however, uterine kinship is never extinguished, and it often happens that people of different clans who are strangers to one another discover that they are distant *saarət*. Sometimes, owing to the devious ramifications of uterine kinship, a married couple discover that they are distant *saarət*. If a child has been born, or if the couple refuse to part for any other reason, their kinship must be ritually severed. This is done by the husband and wife each grasping a calabash with one hand and wrenching it in halves. This symbolizes the splitting of their kinship. The Tallensi think of the uterine line as 'travelling' from one settlement to the next in successive generations, as if it were an entity. Hence a person generally speaks of any uterine clanswoman as 'mother' (*ma*) by identification with his own mother.

The usual range of effective uterine kinship is a good illustration of the correlation between spatial relationship and genealogical relationships. The physical range of the individual's contacts and communication with others is a function of his range of social relations, which is determined principally by his genealogical connections. In the days before the establishment of the British peace uterine kinship played a very important part in enabling people to travel outside their normal range of social contacts, and so to build up social relations with members of clans usually inaccessible to their own clansmen.

Though people can travel where they will in freedom and security nowadays, it is still usual for anyone having business of a ritual or jural kind in a distant or unfamiliar settlement, to follow the web of matrilineal kinship thither. The route may be much curtailed, but the principle remains.

In the same way, uterine kin assist one another in economic matters. A man who loses time in his farming programme through illness will often ask a *soog* from a neighbouring settlement to come and do a day's hoeing for him when he would not ask for such help from a clansman. Again, crops do not ripen at the same time throughout Taleland. It is the custom, at harvest time, especially at the time of the early millet harvest after the hungry months, for people living in areas that have a late harvest to beg (*soh*) gifts of grain from relatives living in early harvesting areas. Later, when the early harvesting settlements have exhausted their supplies, people from these settlements will go to kinsfolk in late harvesting areas to beg for grain to tide them over. Such reciprocal gifts pass between kinsmen of all categories. But the giving is most generous and most cordial between *soog* kin.

In keeping with the unrestricted range of cognatic kinship among the Tallensi, the close ties of kinship between male *saarat* are extended to their children and sometimes to their grandchildren. With men whose fathers were *saarat*, whether they belong to the same clan or not, the formal ties of kinship are often mellowed by the affection and trust of a friendship like that of their fathers. This is rarer with grandsons of male *saarat*, but is not unknown.

3

The Structure of Unilineal Descent Groups[1]

As is well known, Africa has loomed large in British field research in the past twenty-five years. It is, indeed, largely due to the impact of ethnographic data from Africa that British anthropologists are now giving so much attention to social organization, in the widest sense of that term. In this paper what I shall try to do is to sum up some positive contributions that seem to me to have come out of the study of African social organization. I want to add this. British anthropologists are well aware that their range of interests seems narrow in comparison with the wide and adventurous sweep of American anthropology. This has been due to no small extent to lack of numbers and there are signs that a change is on the way with the increase in the number of professional anthropologists since the end of the war. At the same time, I believe that the loss in diversity is amply balanced by the gains we have derived from concentration on a limited set of problems.[2]

Social anthropology has undoubtedly made great progress in the past twenty years. I would give pride of place to the accumulation of ethnographic data obtained by trained observers. It means, curiously enough, that there is going to be more scope than ever for the 'armchair' scholar in framing and testing hypotheses with the help of reliable and detailed information. For Africa the advance from the stage of primitive anecdotage to

[1] This paper (reprinted from *American Anthropologist*, **55**, 1, 1953) was presented at the Symposium on the 'Positive Contributions of Social Anthropology', held at the 50th annual meeting of the American Anthropological Association in Chicago, 15–17 November 1951. My participation in the symposium was made possible by the generosity of the Wenner-Gren Foundation for Anthropological Research Inc.

[2] This was written before I saw the discussion between Dr Murdock and Professor Firth on the limitations of British social anthropology in the *American Anthropologist*, **53**, 4, Pt. 1, 1951.

that of scientific description has been almost spectacular; and most of it has taken place since 1930, as can be judged by comparing what we know today with the state of African ethnography as described by Dr Edwin Smith in 1935. Mainly through Malinowski's influence we now have a respectable series of descriptive monographs on specific institutional complexes in particular African societies. Studies like Evans-Pritchard's on Zande witchcraft (1937), Schapera's on Tswana law (1938) and Richards' on Bemba economy (1939), to cite only three outstanding prewar examples, typify the advance made since 1930. They are significant not only for their wealth of carefully documented detail but also for the evidence they give of the validity of the thesis, now so commonplace, that the customs and institutions of a people can only be properly understood in relation to one another and to the 'culture as a whole'. They show also what a powerful method of ethnographic discovery intensive field work on 'functionalist' lines can be.

The field work of the past two decades has brought into clearer focus the characteristics of African societies which distinguish them from the classical simple societies of, say, Australia, Melanesia or North America; and the mark of this is easily seen in the thought and interests of Africanists. One of these is the relatively great size, in terms both of territorial spread and of numbers, of many ethnographic units in Africa as compared with the classical simple societies. There are few truly isolated societies in Africa. Communication takes place over wide geographical regions; and movements of groups over long stretches of time, exactly like those that are known from our own history, have spread languages, beliefs, customs, craft and food-producing techniques and the network of trade and government, over large areas with big populations. A tribe of ten thousand Tswana, two hundred thousand Bemba or half a million Ashanti cannot run their social life on exactly the same pattern as an Australian horde, which is, after all, basically a domestic group. In Africa one comes up against economics where in Australia or parts of North America one meets only housekeeping; one is confronted with government where in societies of smaller scale one meets social control; with organized warfare, with complex legal institutions, with elaborate forms of public worship and systems of belief comparable to the philosophical and theological systems of literate civilizations.

Even before its subjugation by Europe, Africa boasted big and wealthy towns. Certainly there was knowledge of all this before professional anthropologists began to work in Africa. But it was patchy and on the whole superficial. In particular, it lacked the explicit conceptualization and integral presentation that mark the kind of monograph I have mentioned. That a belief in witchcraft occurred in many African cultures was known long ago. But the precise nature of the belief, and how it was related to the notion of causation, the rules of moral conduct, the practice of divination and the art of healing to form with them a coherent ideology for daily living, was not understood till Evans-Pritchard's book appeared. It was known, from the works of nineteenth-century travellers and administrators, that many African societies had forms of government similar to what political philosophers call the State. But there was little or no accurate information about the constitutional laws, the structure of administration, the machinery of justice, the sanctions of rank, the getting and spending of public revenues, and so forth, in any African state before Rattray's important studies in Ashanti in the twenties (Rattray 1929 and later). Rattray's description of African state structure has now been superseded. We have a pretty good idea of how a monarchy was kept in power not only by ritual constraints and prerogatives, as in the case of the Divine Kingship of the Shilluk (see Evans-Pritchard 1948) but also by means of shrewd secular sanctions and institutions such as the control of public revenues and armed forces in Dahomey, described by Herskovits (1938); or the manipulation of a rank and class based administration as in Nupe (Nadel 1942); or by means of both ritual and secular institutions as has been so vividly described for the Swazi by Dr Hilda Kuper (1947).

Of course, African customs and institutions often have significant resemblances to those of the simpler peoples of other continents. Indeed it is just these resemblances that make the distinctive features of African ethnology stand out in proper theoretical perspective. Take the customs of avoidance between affines or between successive generations, known from many parts of the world. We are apt to think of them, even with reference to such characteristically African cultures as those of the Southern Bantu (cf. Hunter 1936) as expressing specific interpersonal relationships. It is the more striking to find among the

Nyakyusa (Wilson 1951) that the whole scheme of local organization in age villages turns on such avoidances. Moreover we can, in this case, see sharply and writ large, how the avoidance between father-in-law and daughter-in-law is an aspect of the tension between successive generations in a patrilineal kinship system.

Implicit and sometimes explicit comparison of African cultures with those of other areas is important in the recent history of field research in Africa. Seligman's pioneering researches in the Sudan were done against the background of his experiences in New Guinea and among the Veddas (*cf.* C. G. and B. Z. Seligman 1932). More important, though, is the fact that the main theoretical influence behind the field work of British anthropologists in Africa in the middle twenties and the thirties was that of Malinowski. Now Malinowski's 'functional' theory is ordered to the concept of *culture*, essentially in a sense derived from Tylor and Frazer, and his empirical model was always the Trobrianders. It has taken twenty years for the Trobrianders to be placed in a proper comparative perspective in British social anthropology.

It is not, I think, a gross distortion to say that Malinowski thought of culture fundamentally in terms of a utilitarian philosophy. The individual using his culture to satisfy universal needs by attaining culturally defined ends is central to his ethnographic work. It is in the real events of social life, in situations of work, ceremony, dance, dispute, that he saw the interconnection of all aspects of culture. And this approach, crystallized in his formula for the institution – the group, the universal need, the material basis, the legal or mythical charter – has proved to be of the greatest value for the empirical task of field observation. Methodologically, it might be described as a form of clinical study. The net of enquiry is spread to bring in everything that actually happens in the context of observation. The assumption is that everything in a people's culture is meaningful, functional, in the here-and-now of its social existence. This is the cardinal precept for the anthropological study of a living culture. It is the basis of the rigorous observation and comprehensive binding together of detail that marks good ethnographic field work of today. However we may now regard Malinowski's theories we cannot deny him credit for showing us how intensive field work can and must be done. That is, I believe, one of the major contributions made by social anthropology to the social sciences, though it can probably only be satisfactorily used in homogeneous and relatively stable societies or sections of societies.

What I am concerned with in these remarks is the local history of British social anthropology. We all know that Malinowski's functionalism was part of a wider movement; but this is not my subject. The point I am leading up to is this. Malinowski had no sense for social organization, though paradoxically enough his most valuable specific hypotheses fall within the frame of reference of social organization. This applies, for instance to his restatement of the Durkheimian hypothesis of the function of myth as the 'charter' of an institution, to his remarkable analysis of the configuration of social relations in the matrilineal family, and to his development of the concept of reciprocity. But he had no real understanding of kinship or political organization. Thus he never overlooked an opportunity of pouring scorn on what he called 'kinship algebra', as I can vouch for from personal experience. This prejudice prevented him from completing his often promised book on kinship. It is beautifully documented in the *Sexual Life of Savages* (p. 447). Kinship is to him primarily a tissue of culturally conditioned emotional attitudes. So he is puzzled by the extension of the term for 'father' to the father's sister's son: and being quite unable to think in what we should now call structural terms, he commits the appalling methodological solecism of attributing it to an anomaly of language. Malinowski was reacting against the preoccupation with terminologies and with conjectural reconstructions of extinct marriage rules which was so widespread in the early years of this century. It is a measure of the progress made since 1929 that no one today coming across so obvious a case of a Choctaw type lineage terminology would make Malinowski's blunder.

Malinowski's bias is the more instructive because of the debt we owe to his genius. It is reflected in the field work directly inspired by him. We see this in what I regard as the most outstanding contribution to African ethnography we have as yet had, Evans-Pritchard's Zande book (1937). It is notable that he refers only incidentally and casually to the way witchcraft and oracles are tied up with Zande political organization. Firth's study of Tikopia kinship (1936) is an exception for its grasp of the theory of social organization; but he still held the view that social organization is an aspect of culture of the same modality as the others usually enumerated by Malinowski. I mention these two books because they mark important steps in the advance of both ethnography and theory; and I am not suggesting that they follow a wrong track. What I want to stress is that they follow the track which leads to 'culture' as the global concept subsuming everything that goes on in social life. A serious limitation to this point of view is that it is bound to treat everything in social life as of equal weight, all aspects as of equal significance. There is no way of establishing an order of priority where all institutions are interdependent, except by criteria that cannot be used in a synchronic study; and synchronic study is the *sine qua non* of functional research. There is, for instance, the criterion of viability over a stretch of time which enables us to say that parliamentary government is a more vital institution in the British Commonwealth than slavery because it has outlived the latter; or that, for the same reason, matrilineal kinship is more significant among the coastal Akan of the Gold Coast than the worship of their pagan gods. Such a criterion, for what it is worth, is not applicable in the absence of historical documents. It is arguable, of course, that this is a false problem, that in fact all the customs and institutions of a society at a given time *are* of equal weight. But it is not scientifically satisfying to accept this assumption without more ado. If our colleagues in human biology had been content with such an assumption in the nature-nurture problem they would have given up their studies of twins and so left the science of human heredity lacking in some of its most critical data. For human society and culture the problem has hitherto been posed and dogmatically answered by the various brands of determinists. Or at the other extreme it has been implied and subtly evaded by the hypostatization of patterns, geniuses and styles. But the problem remains wide open and Malinowski, in common with all who think in terms of a global concept of culture, had no answer to it.

Social anthropology has made some advance on this position since the thirties. Most social anthropologists would now agree that we cannot, for analytical purposes, deal exhaustively with our ethnographic observations in a single frame of reference. We can regard these observations as facts of custom – as standardized ways of doing, knowing, thinking, and feeling – universally obligatory and valued in a given group of people at a given time. But we can also regard them as facts of social organization or social structure. We then seek to relate them to one another by a scheme of conceptual operations different from that of the previous frame of reference. We see custom as symbolizing or expressing social relations – that is, the ties and cleavages by which persons and groups are bound to one another or divided from one another in the activities of social life. In this sense social structure is not an aspect of culture but the entire culture of a given people handled in a special frame of theory. Lastly, we can consider ethnographic facts in terms of a socio-psychological or bio-psychological frame of reference, seeking relevant connections between them as they come into action in the whole or a part of an individual life process, or more widely, as they represent general human aptitudes and dispositions. And no doubt as our subject develops other special techniques and procedures will emerge for handling the data. No one denies the close connection between the different conceptual frames I have mentioned. By distinguishing them we recognize that different modes of abstraction calling for somewhat different emphases in field enquiry are open to us. What I am saying is commonplace today. It was not so in the middle thirties and this was a source of theoretical weakness as Bateson pointed out (1936).

British anthropologists owe their realization of this methodological distinction both to ethnographic discoveries of recent years and to the catalytic influence exercised on their thought by Radcliffe-Brown since his return to England from Chicago in 1937. But the distinction had of course long been implicit in the work of earlier ethnologists. We need only think of the contrast between Lewis Morgan, whose idiom of thought was in terms of a social system, and Tylor, who thought in terms of custom and often had recourse to psychological hypotheses. Rivers (1914) whose own work and influence in England contributed significantly to the development of the idea of social structure, saw this. So did Lowie whose *Primitive Society* (1921) is, I suppose, the first attempt at a systematic analysis of what we should now call the principles of social structure in primitive society. What he brought out was the very obvious but fundamental fact that closely similar, if not identical, forms of social relationship occur in widely separate societies and are expressed in varied custom.

By social organization or social structure, terms which they used interchangeably, Rivers and Lowie meant primarily the kinship, political and legal institutions of primitive peoples. And these, in fact, are the institutions with which British anthropologists are mainly concerned when they write about social structure. The advantage of this term, as opposed to the more usual term 'social organization' is that it draws attention to the interconnection and interdependence, within a single system of all the different classes of social relations found within a given society. This leads to questions being asked about the nature of these interconnections and the forces behind the system as a whole.

What I want to stress is that the spur to the current interest in structural studies in Britain comes in equal measure from field experience, especially in Africa, and theory. Anybody who has tried to understand African religious beliefs and practices in the field knows, for example, that one cannot get far without a very thorough knowledge of the kinship and political organization. These studies have thus given new content to the familiar postulate that a living culture is an integrated unity of some sort. We can see more clearly than twenty years ago that this is due not to metaphysical qualities mysteriously diffused through it but to the function of customs and institutions in expressing, marking and maintaining social relations between persons and groups. It is this which underlies the consistencies between the customs and institutions of a people that are commonly emphasized. A unit must, by definition, have a boundary. A culture, certainly in most of Africa, and I venture to believe in many other areas too (as indeed Wissler long ago stressed), has no clear-cut boundaries. But a group of people bound together within a single social structure have a boundary, though not necessarily one that coincides with a physical boundary or is impenetrable. I would suggest that a culture is a unity in so far as it is tied to a bounded social structure. In this sense I would agree that the social structure is the foundation of the whole social life of any *continuing* society. Here again Rivers showed great insight when he stated (1911) that the social structure is the feature of a people's social life which is most resistant to change. It is certainly a striking fact that the family and kinship institutions of a continuing society in Africa display remarkable persistence in the face of big changes in everyday habits, in ritual customs and belief, and even in major economic and social goals. The Tswana (*cf.* Schapera 1940 and 1950) are a good instance. But we must be careful. There is also plenty of evidence from emigrant groups, such as Chinese, East Indians and particularly the Negro populations of the New World (*cf.* Herskovits 1949, pp. 542 ff) of the retention of religious and aesthetic customs in the face of radical changes in structural arrangements. This is a warning against thinking of culture and social structure as mutually exclusive. The social structure of a group does not exist without the customary norms and activities which work through it. We might safely conclude that where structure persists there must be some persistence of corresponding custom and where custom survives there must be some structural basis for this. But I think it would be agreed that though the customs of any continuing and stable society tend to be consistent because they are tied to a coherent social structure, yet there are important factors of autonomy in custom. This has often been pointed out ever since the facts of diffusion became known. The part played by dispositional and psychogenetic factors in the content and action of custom is now being clarified. A house is not reducible to its foundations and custom is not reducible simply to a manifestation of social structure.

The recent trend in British social anthropology springs, as I have said, primarily from field experience. Evans-Pritchard's description of Nuer lineage organization (1933–5), Firth's account of Tikopia kinship (1936) and Forde's analysis of clan and kin relations among the Yakö (1938–9) are the important ethnographic landmarks. A prominent feature in all three is the attention given to the part played by descent rules and institutions in social organization, and the recognition that they belong as much to the sphere of political organization as to that of kinship. Following this lead, other students have been making intensive studies of the role of descent principles in African societies where unilineal descent groups often constitute the genealogical basis of social relations. Good ethnography is both a continuous test of existing hypotheses and continuously creative of theory and technique; and this is happening so rapidly just at present that one can hardly keep pace with it. The younger research workers to whose unpublished material I shall be referring are developing structural analysis into a very effective technique and applying it not only in Africa but also in India, New Guinea and Indonesia.

Seen against the background I have sketched, there is no doubt that big gains have been made in the study of social structure since the nineteen-twenties. This is well illustrated in recent investigations of unilineal descent groups, both in Africa and elsewhere (cf. Eggan 1950; Gough 1950) but I will deal mainly with the African data. We are now in a position to formulate a number of connected generalizations about the structure of the unilineal descent group, and its place in the total social system which could not have been stated twenty years ago. It is moreover important to note that they seem to hold for both patrilineal and matrilineal groups. Some of the conditions governing the emergence of such descent groups have recently been discussed by Forde (1947). He makes the interesting suggestion that poverty of habitat and of productive technology tend to inhibit the development of unilineal descent groups by limiting the scale and stability of settlement. Taking this in association with Lowie's hypothesis of 1921 (Lowie 1921, p. 149) that the establishment of the principle of unilateral descent is mainly due to the transmission of property rights and the mode of residence after marriage, we have two sides of an hypothesis that deserves much further testing. The ground has been well cleared for this by Murdock (1949). For it does seem that unilineal descent groups are not of significance among peoples who live in small groups, depend on a rudimentary technology, and have little durable property. On the other hand, there is evidence that they break down when a modern economic framework with occupational differentiation linked to a wide range of specialized skills, to productive capital and to monetary media of exchange is introduced (Spoehr 1947; Eggan 1950; Gough 1950). Where these groups are most in evidence is in the middle range of relatively homogeneous, pre-capitalistic economies in which there is some degree of technological sophistication and value is attached to rights in durable property. They may be pastoral economies like the Nuer (Evans-Pritchard 1940) and the Beduin (Peters 1951), or agricultural economies like those of the Yakö (Forde 1938, 1950), the Tallensi (Fortes 1945, 1949) and the Gusii (Mayer 1949) – or if we look outside Africa, the Tikopia (Firth

1936) and the Hopi (Eggan 1950) and many other peoples. The Nayar of South India, classically a test case of kinship theories, are of particular interest in this connection, as a recent intensive field study by Dr E. J. Miller and Dr E. K. Gough shows. Though the total economy of South India was even formerly a very complex one, the Nayar themselves traditionally formed a caste of very limited occupational range. It is only during the past hundred years or so that they have gradually entered other occupations than soldiering and passive landlordism. And with this change has come the breakdown previously mentioned in their rigid matrilineal lineage organization. This does not imply that unilineal descent groups are either historically or functionally the product of economic and property institutions alone. Other factors are undoubtedly involved. There is the example of the Hausa of Northern Nigeria, for instance, who have a rural economy of the same type as that of the Tallensi, though technically more elaborate, and well developed property concepts; but they have no unilineal descent groups. The socially significant genealogical grouping among them is of the cognatic type based on the equal recognition of kin ties on both sides, as among the Lozi and other Central African tribes (Dry 1950; Colson and Gluckman 1951). Nor can the Hausa arrangement be ascribed to the local influence of Islam since the Cyrenaican Beduin have sharply defined patrilineal lineages (Peters 1951).

I have lingered a little on this problem to bring home a point which I have already referred to. It is the problem of assigning an order of relative weight to the various factors involved in culture and in social organization, or alternatively, of devising methods for describing and analysing a configuration of factors so as to show precisely how they interact with one another. Much as we have learned from intensive field work in relation to this task, we shall learn even more, I believe, from such studies of local variations within a uniform culture region as Radcliffe-Brown's (1930), Schapera's (in Radcliffe-Brown and Forde 1950) and Eggan's (1950).

The most important feature of unilineal descent groups in Africa brought into focus by recent field research is their corporate organization. When we speak of these groups as corporate units we do so in the sense given to the term 'corporation' long ago by Maine in his classical analysis of testamentary succession in early law (Maine 1861). We are reminded also of Max Weber's sociological analysis of the corporate group as a general type of social formation (Weber 1947), for in many important particulars these African descent groups conform to Weber's definition. British anthropologists now regularly use the term *lineage* for these descent groups. This helps both to stress the significance of descent in their structure and to distinguish them from wider often dispersed divisions of society ordered to the notion of common – but not demonstrable and often mythological – ancestry for which we find it useful to reserve the label *clan*.

The guiding ideas in the analysis of African lineage organization have come mainly from Radcliffe-Brown's formulation of the structural principles found in all kinship systems (*cf.* Radcliffe-Brown 1950). I am sure I am not alone in regarding these as among the most important generalizations as yet reached in the study of social structure. Lineage organization shows very clearly how these principles work together in mutual dependence, so that varying weight of one or the other in relation to variations in the wider context of social structure gives rise to variant arrangements on the basis of the same broad ground-plan.

A lineage is a corporate group from the outside, that is in relation to other defined groups and associations. It might be described as a single legal personality – 'one person' as the Ashanti put it (Fortes 1950). Thus the way a lineage system works depends on the kind of legal institutions found in the society; and this, we know, is a function of its political organization. Much fruitful work has resulted from following up this line of thought. As far as Africa is concerned there is increasing evidence to suggest that lineage organization is most developed in what Evans-Pritchard and I (1940), taking a hint from Durkheim, called segmentary societies. This has been found to hold for the Tiv of Nigeria (P. J. Bohannan 1951), for the Gusii (Mayer 1949) and other East and South African peoples, and for the Cyrenaican Beduin (Peters 1951), in addition to the peoples discussed in *African Political Systems*. In societies of this type the lineage is not only a corporate unit in the legal or jural sense but is also the primary political association. Thus the individual has no legal or political status except as a member of a lineage; or to put it in another way, all legal and political relations in the society take place in the context of the lineage system.

But lineage grouping is not restricted to segmentary societies. It is the basis of local organization and of political institutions also in societies like the Ashanti (Fortes 1950; Busia 1951) and the Yoruba (Forde 1951) which have national government centred in kingship, administrative machinery and courts of law. But the primary emphasis, in these societies, is on the legal aspect of the lineage. The political structure of these societies was always unstable and this was due in considerable degree to internal rivalries arising out of the divisions between lineages; that is perhaps why they remained federal in constitution. In Ashanti, for instance, this is epitomized in the fact that citizenship is, in the first place, local not national, is determined by lineage membership by birth and is mediated through the lineage organization. The more centralized the political system the greater the tendency seems to be for the corporate strength of descent groups to be reduced or for such corporate groups to be nonexistent. Legal and political status are conferred by allegiance to the State not by descent, though rank and property may still be vested in descent lines. The Nupe (Nadel 1942), the Zulu (Gluckman in Fortes and Evans-Pritchard 1940), the Hausa (Dry 1950), and other state organizations exemplify this in different ways. There is, in these societies, a clearer structural differentiation between the field of domestic relations based on kinship and descent and the field of political relations, than in segmentary societies.

However, where the lineage is found as a corporate group all the members of a lineage are to outsiders jurally equal and represent the lineage when they exercise legal and political rights and duties in relation to society at large. This is what underlies so-called collective responsibility in blood vengeance and self-help as among the Nuer (Evans-Pritchard 1940) and the Beduin (Peters 1951).

Maine's aphorism that corporations never die draws attention to an important characteristic of the lineage, its continuity, or rather its presumed perpetuity in time. Where the lineage concept is highly developed, the lineage is thought to exist as a perpetual corporation as long as any of its members survive. This means, of course, not merely perpetual physical existence ensured by the replacement of departed members. It means perpetual structural existence, in a stable and homogeneous society; that is, the perpetual exercise of defined rights, duties, office and social tasks vested in the lineage as a corporate unit. The point is obvious but needs recalling as it throws light on a widespread custom. We often find, in Africa and elsewhere, that a person or descent group is attached to a patrilineal lineage through a female member of the lineage. Then if there is a danger that rights and offices vested in the lineage may lapse through the extinction of the true line of descent, the attached line may by some jural fiction be permitted to assume them. Or again, rather than let property or office go to another lineage by default of proper succession within the owning lineage, a slave may be allowed to succeed. In short, the aim is to preserve the existing scheme of social relations as far as possible. As I shall mention presently, this idea is developed most explicitly among some Central African peoples.

But what marks a lineage out and maintains its identity in the face of the continuous replacement by death and birth of its members is the fact that it emerges most precisely in a complementary relationship with or in opposition to like units. This was first precisely shown for the Nuer by Evans-Pritchard and I was able to confirm the analysis among the Tallensi (Fortes 1949a). It is characteristic of all segmentary societies in Africa so far described, almost by definition. A recent and most interesting case is that of the Tiv of Northern Nigeria (P. J. Bohannan 1951). This people were, until the arrival of the British, extending their territory rapidly by moving forward *en masse* as their land became exhausted. Among them the maximal lineages are identified by their relative *positions* in the total deployment of all the lineages and they maintain these positions by pushing against one another as they all move slowly forward.

The presumed perpetuity of the lineage is what lineage genealogies conceptualize. If there is one thing all recent investigations are agreed upon it is that lineage genealogies are not historically accurate. But they can be understood if they are seen to be the conceptualization of the existing lineage structure viewed as continuing through time and therefore projected backward as pseudo-history. The most striking proof of this comes from Cyrenaica. The Beduin there have tribal genealogies going back no more than the fourteen generations or thereabouts which we so commonly find among African Negro peoples; but as Peters points out, historical records show that they have lived in Cyrenaica apparently in much the same way as now for a much longer

time than the four to five hundred years implied in their genealogies. Dr P. J. and Dr L. Bohannan have actually observed the
Tiv at public moots rearranging their lineage genealogies to
bring them into line with changes in the existing pattern of legal
and political relations within and between lineages. A genealogy
is, in fact, what Malinowski called a legal charter and not an
historical record.

A society made up of corporate lineages is in danger of splitting
into rival lineage factions. How is this counteracted in the interests
of wider political unity? One way is to extend the lineage framework to the widest range within which sanctions exist for pre-
venting conflicts and disputes from ending in feud or warfare.
The political unit is thought of then as the most inclusive, or
maximal, lineage to which a person can belong, and it may be
conceptualized as embracing the whole tribal unit. This happens
among the Gusii (Mayer 1949) as well as among the Nuer, the
Tiv and the Beduin; but with the last three the tribe is not the
widest field within which sanctions against feud and war prevail.
A major lineage segment of the tribe is the *de facto* political unit
by this definition.

Another way, widespread in West Africa but often associated
with the previously mentioned structural arrangement, is for the
common interest of the political community to be asserted
periodically, as against the private interests of the component
lineages, through religious institutions and sanctions. I found
this to be the case among the Tallensi (Fortes 1940) and the same
principle applies to the Yakö (Forde 1950b) and the Ibo (Forde
and Jones 1950). I believe it will be shown to hold for many
peoples of the Western Sudan among whom ancestor worship
and the veneration of the earth are the basis of religious custom.
The politically integrative functions of ritual institutions have
been described for many parts of the world. What recent African
ethnography adds is detailed descriptive data from which further
insight into the symbolism used and into the reasons why political
authority tends to be invested with ritual meaning and expression
can be gained. A notable instance is Dr Kuper's (1947) account
of the Swazi kingship.

As the Swazi data indicate, ritual institutions are also used to
support political authority and to affirm the highest common
interests in African societies with more complex political

structures than those of segmentary societies. This has long been known, ever since the Divine Kingship of the Shilluk (*cf.* Evans-Pritchard 1948) brought inspiration to Sir James Frazer. But these ritual institutions do not free the individual to have friendly and co-operative relations with other individuals irrespective of allegiance to corproate groups. If such relations were impossible in a society it could hardly avoid splitting into antagonistic factions in spite of public ritual sanctions, or else it would be in a chronic state of factional conflict under the surface. It is not surprising therefore to find that great value is attached to widely spreading bonds of personal kinship, as among the Tallensi (Fortes 1949a). The recent field studies I have quoted all confirm the tremendous importance of the web of kinship as a counter-weight to the tendency of unilineal descent grouping to harden social barriers. Or to put it slightly differently, it seems that where the unilineal descent group is rigorously structured within the total social system that we are likely to find kinship used to define and sanction a personal field of social relations for each individual. I will come back to this point in a moment. A further point to which I will refer again is this. We are learning from considerations such as those I have just mentioned, to think of social structure in terms of levels of organization in the manner first explicitly followed in the presentation of field data by Warner (1937). We can investigate the total social structure of a given community at the level of local organization, at that of kinship, at the level of corporate group structure and government, and at that of ritual institutions. We see these levels are related to different collective interests, which are perhaps connected in some sort of hierarchy. And one of the problems of analysis and exposition is to perceive and state the fact that all levels of structure are simultaneously involved in every social relationship and activity. This restatement of what is commonly meant by the concept of integration has the advantage of suggesting how the different modes of social relationship distinguished in any society are interlocked with one another. It helps to make clear also how certain basic principles of social organization can be generalized throughout the whole structure of a primitive society, as for instance the segmentary principle among the Nuer and the Tallensi.

This way of thinking about the problem of social integration has been useful in recent studies of African political organization. Study of the unilineal descent group as a part of a total social system means in fact studying its functions in the widest framework of social structure, that of the political organization. A common and perhaps general feature of political organization in Africa is that it is built up in a series of layers, so to speak, so arranged that the principle of checks and balances is necessarily mobilized in political activities. The idea is used in a variety of ways but what it comes to in general is that the members of the society are distributed in different, nonidentical schemes of allegiance and mutual dependence in relation to administrative, juridical and ritual institutions. It would take too long to enumerate all the peoples for whom we now have sufficient data to show this in detail. But the Lozi of Northern Rhodesia (Gluckman 1951) are of such particular theoretical interest in this connection that a word must be said about them. The corporate descent group is not found among them. Instead their political organization is based on what Maine called the corporation sole. This is a title carrying political office backed by ritual sanctions and symbols to which subjects, lands, jurisdiction, and representative status, belong. But every adult is bound to a number of titles for different legal and social purposes in such a way that what is one allegiance group with respect to one title is split up with reference to other titles. Thus the only all-inclusive allegiance is that of all the nation to the kingship, which is identified with the State and the country as a whole. A social structure of such a kind, knit together moreover by a widely ramifying network of bilateral kinship ties between persons, is well fortified against internal disruption. It should be added that the notion of the 'corporation sole' is found among many Central African peoples. It appears, in fact, to be a jural institution of the same generality in any of these societies as corporate groups are in others, since it is significant at all levels of social structure. A good example is the Bemba (*cf.* Richards 1934, 1940b) among whom it is seen in the custom of 'positional inheritance' of status, rank, political office and ritual duty, as I will explain later.

What is the main methodological contribution of these studies? In my view it is the approach from the angle of political organization to what are traditionally thought of as kinship groups and institutions that has been specially fruitful. By regarding lineages and statuses from the point of view of the total social system and not from that of an hypothetical EGO we realize that consanguinity and affinity, real or putative, are not sufficient in themselves to bring about these structural arrangements. We see that descent is fundamentally a jural concept as Radcliffe-Brown argued in one of his most important papers (1935a); we see its significance, as the connecting link between the external, that is political and legal, aspect of what we have called unilineal descent groups, and the internal or domestic aspect. It is in the latter context that kinship carries maximum weight, first, as the source of title to membership of the groups or to specific jural status, with all that this means in rights over and towards persons and property, and second as the basis of the social relations among the persons who are identified with one another in the corporate group. In theory, membership of a corporate legal or political group need not stem from kinship, as Weber has made clear. In primitive society, however, if it is not based on kinship it seems generally to presume some formal procedure of incorporation with ritual initiation. So-called secret societies in West Africa seem to be corporate organizations of this nature. Why descent rather than locality or some other principle forms the basis of these corporate groups is a question that needs more study. It will be remembered that Radcliffe-Brown (1935a) related succession rules to the need for unequivocal discrimination of rights *in rem* and *in personam*. Perhaps it is most closely connected with the fact that rights over the reproductive powers of women are easily regulated by a descent group system. But I believe that something deeper than this is involved; for in a homogeneous society there is nothing which could so precisely and incontrovertibly fix one's place in society as one's parentage.

Looking at it from without, we ignore the internal structure of the unilineal group. But African lineages are not monolithic units; and knowledge of their internal differentiation has been much advanced by the researches I have mentioned. The dynamic character of lineage structure can be seen most easily in the balance that is reached between its external relations and its internal structure. Ideally, in most lineage-based societies the lineage tends to be thought of as a perpetual unit, expanding like a balloon but never growing new parts. In fact, of course, as Forde (1938) and Evans-Pritchard (1940) have so clearly shown, fission and accretion are processes inherent in lineage structure. However, it is a common experience to find an informant who refuses to admit that his lineage or even his branch of a greater lineage did not at one time exist. Myth and legend, believed, naturally, to be true history, are quickly cited to prove the contrary. But investigation shows that the stretch of time, or rather of duration, with which perpetuity is equated varies according to the count of generations needed to conceptualize the internal structure of the lineage and link it on to an absolute, usually mythological origin for the whole social system in a first founder.

This is connected with the fact than an African lineage is never, according to our present knowledge, internally undifferentiated. It is always segmented and is in process of continuous further segmentation at any given time. Among some of the peoples I have mentioned (e.g. the Tallensi and probably the Ibo) the internal segmentation of a lineage is quite rigorous and the process of further segmentation has an almost mechanical precision. The general rule is that every segment is, in form, a replica of every other segment and of the whole lineage. But the segments are, as a rule, hierarchically organized by fixed steps of greater and greater inclusiveness, each step being defined by genealogical reference. It is perhaps hardly necessary to mention again that when we talk of lineage structure we are really concerned, from a particular analytical angle, with the organization of jural, economic, and ritual activities. The point here is that lineage segmentation corresponds to gradation in the institutional norms and activities in which the total lineage organization is actualized. So we find that the greater the time depth that is attributed to the lineage system as a whole, the more elaborate is its internal segmentation. As I have already mentioned, lineage systems in Africa, when most elaborate, seem to have a maximal time depth of around fourteen putative generations. More common though is a count of five or six generations of named ancestors between living adults and a quasi-mythological founder. We can as yet only guess at the conditions that lie behind these limits of genealogical depth in lineage structure. The facts themselves are nevertheless of great comparative interest. As I have previously remarked, these genealogies obviously do not represent a true record of all the ancestors of a group. To explain this by the limitations and fallibility of oral tradition is merely to evade the problem. In structural terms the answer seems to lie in the spread or span (Fortes 1945) of internal segmentation of the lineage, and this apparently has inherent limits. As I interpret the evidence we have, these limits are set by the condition of stability in the social structure which it is one of the chief functions of lineage systems to maintain. The segmentary spread found in a given lineage system is that which makes for maximum stability; and in a stable social system it is kept at a particular spread by continual internal adjustments which are conceptualized by clipping, patching and telescoping genealogies to fit. Just what the optimum spread of lineage segmentation in a particular society tends to be depends presumably on extra-lineage factors of political and economic organization of the kind referred to by Forde (1947).

It is when we consider the lineage from within that kinship becomes decisive. For lineage segmentation follows a model laid down in the parental family. It is indeed generally thought of as the perpetuation, through the rule of the jural unity of the descent line and of the sibling group (*cf.* Radcliffe-Brown 1950), of the social relations that constitute the parental family. So we find a lineage segment conceptualized as a sibling group in symmetrical relationship with segments of a like order. It will be a paternal sibling group where descent is patrilineal and a maternal one where it is matrilineal. Progressive orders of inclusiveness are formulated as a succession of generations; and the actual process of segmentation is seen as the equivalent of the division between siblings in the parental family. With this goes the use of kinship terminology and the application of kinship norms in the regulation of intra-lineage affairs.

As a corporate group, a lineage exhibits a structure of authority, and it is obvious from what I have said why this is aligned with the generation ladder. We find, as a general rule, that not only the lineage but also every segment of it has a head, by succession or election, who manages its affairs with the advice of his co-members. He may not have legal sanction by means of which to enforce his authority in internal affairs; but he holds his position by consent of all his fellow members, and he is backed by moral sanctions commonly couched in religious concepts. He is the trustee for the whole group of the property and other productive resources vested in it. He has a decisive jural role also in the disposal of rights over the fertility of the women in the group. He is likely to be the representative of the whole group in political and legal relations with other groups, with political authorities, and in communal ritual. The effect may be to make him put the interests of his lineage above those of the community if there is conflict with the latter. This is quite clearly recognized by some peoples. Among the Ashanti for instance, every chiefship is vested in a matrilineal lineage. But once a chief has been installed his constitutional position is defined as holding an office that belongs to the whole community not to any one lineage. The man is, ideally, so merged in the office that he virtually ceases to be a member of his lineage, which always has an independent head for its corporate affairs (*cf.* Busia 1951).

Thus lineage segmentation as a process in time links the lineage with the parental family; for it is through the family that the lineage (and therefore the society) is replenished by successive generations; and it is on the basis of the ties and cleavages between husband and wife, between polygynous wives, between siblings, and between generations that growth and segmentation take place in the lineage. Study of this process has added much to our understanding of well known aspects of family and kinship structure.

I suppose that we all now take it for granted that filiation – by contrast with descent – is universally bilateral. But we have also been taught, perhaps most graphically by Malinowski, that this does not imply equality of social weighting for the two sides of kin connection. Correctly stated, the rule should read that filiation is always complementary, unless the husband in a matrilineal society (like the Nayar) or the wife in a patrilineal society, as perhaps in ancient Rome, is given no parental status or is legally severed from his or her kin. The latter is the usual situation of a slave spouse in Africa.

Complementary filiation appears to be the principal mechanism by which segmentation in the lineage is brought about. This is very clear in patrilineal descent groups, and has been found to hold for societies as far apart as the Tallensi in West Africa and the Gusii in East Africa. What is a single lineage in relation to a male founder is divided into segments of a lower order by reference to their respective female founders on the model of the division of a polygynous family into separate matricentral 'houses.' In matrilineal lineage systems, however, the position is different. Segmentation does not follow the lines of different paternal origin, for obvious reasons; it follows the lines of differentiation between sisters. There is a connection between this and the weakness in law and in sentiment of the marriage tie in matrilineal societies, though it is usual for political and legal power to be vested in men as Kroeber (1938) and others have remarked. More study of this problem is needed.

Since the bilateral family is the focal element in the web of kinship, complementary filiation provides the essential link between a sibling group and the kin of the parent who does not determine descent. So a sibling group is not merely differentiated within a lineage but is further distinguished by reference to its kin ties outside the corporate unit. This structural device allows of degrees of individuation depending on the extent to which filiation on the non-corporate side is elaborated. The Tiv, for example, recognize five degrees of matrilateral filiation by which a sibling group is linked with lineages other than its own. These and other ties of a similar nature arising out of marriage exchanges result in a complex scheme of individuation for distinguishing both sibling groups and persons within a single lineage (L. Bohannan 1951). This, of course, is not unique and has long been recognized, as everyone familiar with Australian kinship systems knows. Its more general significance can be brought out however by an example. A Tiv may claim to be living with a particular group of relatives for purely personal reasons of convenience or affection. Investigation shows that he has in fact made a choice of where to live within a strictly limited range of nonlineage kin. What purports to be a voluntary act freely motivated in fact presupposes a structural scheme of individuation. This is one of the instances which show how it is possible and feasible to move from the structural frame of reference to another, here that of the social psychologist, without confusing data and aims.

Most far-reaching in its effects on lineage structure is the use of the rule of complementary filiation to build double unilineal systems and some striking instances of this are found in Africa. One of the most developed systems of this type is that of the Yakö; and Forde's excellent analysis of how this works (Forde 1950) shows that it is much more than a device for classifying kin. It is a principle of social organization that enters into all social relations and is expressed in all important institutions. There is the division of property, for instance, into the kind that is tied to the patrilineal lineage and the kind that passes to matrilineal kin. The division is between fixed and, in theory, perpetual productive resources, in this case farm land, with which goes residence rights, on the one hand, and on the other, movable and consumable property like livestock and cash. There is a similar polarity in religious cult and in the political office and authority linked with cult, the legally somewhat weaker matrilineal line being ritually somewhat stronger than the patrilineal line. This balance between ritual and secular control is extended to the fertility of the women. An analogous double descent system has been described for some Nuba Hill tribes by Nadel (1950) and its occurrence among the Herero is now classical in ethnology. The arrangement works the other way round, too, in Africa, as among the Ashanti, though in their case the balance is far more heavily weighted on the side of the matrilineal lineage than on that of the jurally inferior and noncorporate paternal line.

These and other instances lead to the generalization that complementary filiation is not merely a constant element in the pattern of family relationships but comes into action at all levels of social structure in African societies. It appears that there is a tendency for interests, rights and loyalties to be divided on broadly complementary lines, into those that have the sanction of law or other public institutions for the enforcement of good conduct, and those that rely on religion, morality, conscience and sentiment for due observance. Where corporate descent groups exist the former seem to be generally tied to the descent group, the latter to the complementary line of filiation.

If we ask where this principle of social structure springs from we must look to the tensions inherent in the structure of the parental family. These tensions are the result of the direction given to individual lives by the total social structure but they also provide the models for the working of that structure. We now have plenty of evidence to show how the tensions that seem normally to arise between spouses, between successive generations and between siblings find expression in custom and belief. In a homogeneous society they are apt to be generalized over wide areas of the social structure. They then evoke controls like the Nyakyusa separation of successive generations of males in age villages that are built into the total social structure by the device of handing over political power to each successive generation as it reaches maturity (Wilson 1951). Or this problem may be dealt with on the level of ritual and moral symbolism by separating parent and first born child of the same sex by taboos that eliminate open rivalry, as among the Tallensi, the Nuer, the Hausa and other peoples.

Thus by viewing the descent group as a continuing process through time we see how it binds the parental family, its growing point, by a series of steps into the widest framework of social structure. This enables us to visualize a social system as an integrated unity at a given time and over a stretch of time in relation to the process of social reproduction and in a more rigorous way than does a global concept of culture.

I do want to make clear, though, that we do not think of a lineage as being just a collection of people held together by the accident of birth. A descent group is an arrangement of persons that serves the attainment of legitimate social and personal ends. These include the gaining of a livelihood, the setting up of a family and the preservation of health and well-being as among the most important. I have several times remarked on the connection generally found between lineage structure and the ownership of the most valued productive property of the society, whether it be land or cattle or even the monopoly of a craft like blacksmithing. It is of great interest, for instance, to find Dr Richards attributing the absence of a lineage organization among the Bemba to their lack of heritable right in land or livestock (Richards 1950). A similar connection is found between lineage organization and the control over reproductive resources and relations as is evident from the common occurrence of exogamy as a criterion of lineage differentiation. And since citizenship is derived from lineage membership and legal status depends on it, political and religious office of necessity vests in lineages. We must expect to find and we do find that the most important religious and magical concepts and institutions of a lineage based society are tied into the lineage structure serving both as the necessary symbolical representation of the social system and as its regulating values. This is a complicated subject about which much more needs to be known. Cults of gods and of ancestors, beliefs of a totemic nature, and purely magical customs and practices, some or all are associated with lineage organization among the peoples previously quoted. What appears to happen is that every significant structural differentiation has its specific ritual symbolism, so that one can, as it were, read off from the scheme of ritual differentiation the pattern of structural differentiation and the configuration of norms of conduct that goes with it. There is, to put it simply, a segmentation of ritual allegiance corresponding to the segmentation of genealogical grouping. Locality, filiation, descent, individuation, are thus symbolized.

Reference to locality reminds us of Kroeber's careful argument of 1938 in favour of the priority of the local relationships of residence over those of descent in determining the line that is legally superior. A lineage cannot easily act as a corporate group if its members can never get together for the conduct of their affairs. It is not surprising therefore to find that the lineage in African societies is generally locally anchored; but it is not necessarily territorially compact or exclusive. A compact nucleus may be enough to act as the local centre for a group that is widely dispersed. I think it would be agreed that lineage and locality are independently variable and how they interact depends on other factors in the social structure. As I interpret the evidence, local ties are of secondary significance, *pace* Kroeber, for local ties do not appear to give rise to structural bonds in and of themselves. There must be common political or kinship or economic or ritual interests for structural bonds to emerge. Again spatial dispersion does not immediately put an end to lineage ties or to the ramifying kin ties found in cognatic systems like that of the Lozi. For legal status, property, office and cult act centripetally to hold dispersed lineages together and to bind scattered kindred. This is important in the dynamic pattern of lineage organization for it contains within itself the springs of disintegration, at the corporate level in the rule of segmentation, at the individual level in the rule of complementary filiation.

As I have suggested before, it seems that corporate descent groups can exist only in more or less homogeneous societies. Just what we mean by a homogeneous society is still rather vague though we all use the term lavishly. The working definition I make use of is that a homogeneous society is ideally one in which any person in the sense given to this term by Radcliffe-Brown in his recent (1950) essay, can be substituted for any other person of the same category without bringing about changes in the social structure. This implies that any two persons of the same category have the same body of customary usages and beliefs. I relate this tentative definition to the rule of sibling equivalence, so that I would say that, considered with respect to their achievable life histories, in a homogeneous society all men are brothers and all women sisters.

Societies based on unilineal descent groups are not the best in which to see what the notion of social substitutability means. For that it is better to consider societies in which descent still takes primacy over all other criteria of association and classification of persons in the regulation of social life but does not serve as the constitutive principle of corporate group organization. Central Africa provides some admirable instances (*cf.* Richards 1950; Colson and Gluckman 1951). Among the Bemba, the Tonga, the Lozi and many of their neighbours, as I have already remarked, the social structure must be thought of as a system of interconnected politico-legal statuses symbolized and sanctioned by ritual and not as a collection of people organized in self-perpetuating descent units. The stability of the society over time is preserved by perpetuating the status system. Thus when a person dies his status is kept alive by being taken up by an heir; and this heir is selected on the basis of descent rules. At any given time an individual may be the holder of a cluster of statuses; but these may be distributed among several persons on his death in a manner analogous to the widespread African custom by which a man's inherited estate goes to his lineage heir and his self-acquired property to his personal heir. Ideally, therefore, the network of statuses remains stable and perpetual though their holders come and go. Ritual symbols define and sanction the key positions in the system. What it represents, in fact, is the generalization throughout a whole society of the notion of the corporation sole as tied to descent but not to a corporate group. Descent and filiation have the function of selecting individuals for social positions and roles – in other words, for the exercise of particular rights and obligations – just as in cross cousin marriage they serve to select ego's spouse.

The concept of the 'person' as an assemblage of statuses has been the starting point of some interesting enquiries. A generalization of long standing is that a married person always has two mutually antagonistic kinship statuses, that of spouse and parent in one family context and that of child and sibling in another (*cf.* Warner 1937). This is very conspicuous in an exogamous lineage system; and the tensions resulting from this condition, connected as they are with the rule of complementary filiation, have wide consequences. A common rule of social structure reflected in avoidance customs is that these two statuses must not be confounded. Furthermore, each status can be regarded as a compound of separable rights and obligations. Thus a problem that has to be solved in every matrilineal society is how to reconcile the rights over a woman's procreative powers (rights *in genetricem* as Laura Bohannan has called them in her paper of 1949) which remain vested in her brother or her lineage, with those over her domestic and sexual services (rights *in uxorem*, *cf.* L. Bohannan, *loc. cit.*) which pass to her husband. Among the Yao of Nyasaland, as Dr Clyde Mitchell has shown (1950), this problem underlies the process of lineage segmentation. Brothers struggle against one another (or sisters' sons against mothers' brothers) for the control of their sisters' procreative powers and this leads to fission in the minimal lineage. It is of great significance that such a split is commonly precipitated by accusations of witchcraft against the brother from whose control the sisters are withdrawn. By contrast, where rights over a woman's childbearing powers are held by her husband's patrilineal lineage the conflicts related to this critical interest occur between the wives of a lineage segment; and among the Zulu- and Xhosa-speaking tribes of South Africa these lead to witchcraft accusations between co-wives (*cf.* Hunter 1936). As Laura Bohannan's paper shows, many widespread customs and institutions connected with marriage and parenthood, such as the levirate and the sororate, wife-taking by women, exchange marriage as practised by the Tiv, and ghost marriage as found among the Nuer (Evans-Pritchard 1951a) have structural significance not hitherto appreciated if they are regarded from the point of view I have indicated.

But one thing must be emphasized. This method of analysis does not explain why in one society certain kinds of interpersonal conflict are socially projected in witchcraft beliefs whereas in another they may be projected in terms of a belief in punitive spirits. It makes clear why a funeral ceremony is necessary and why it is organized in a particular way in the interest of maintaining a stable and coherent social system. It does not explain why the ritual performed in the funeral ceremonies of one people uses materials, ideas and dramatizations of a different kind from those used by another people. In short, it brings us nearer than we were thirty years ago to understanding the machinery by which norms are made effective, not only in a particular primitive society but in a type of primitive society. It does not explain how the norms come to be what they in fact are in a particular society.

In this connection, however, it is worth drawing attention to certain norms that have long been recognized to have a critical value in social organization. Marriage regulations, incest prohibitions and the laws of homicide and warfare are the most important. Analysis of lineage structure has revealed an aspect of these norms which is of great theoretical interest. It is now fairly evident that these are not absolute rules of conduct which men are apt to break through an outburst of unruly instinct or rebellious self-assertion, as has commonly been thought. They are *relatively* obligatory in accordance with the structural relations of the parties. The Beduin of Cyrenaica regard homicide within the minimal agnatic lineage, even under extreme provocation, as a grave sin, whereas slaying a member of a different tribal segment is an admirable deed of valour. The Tallensi consider sex relations with a near sister of the same lineage as incest but tacitly ignore the act if the parties are very distant lineage kin. Among the Tiv, the Nuer, the Gusii and other tribes the lineage range within which the rule of exogamy holds is variable and can be changed by a ceremony that makes formally prohibited marriages legitimate and so brings marriage prohibitions into line with changes in the segmentary structure of the lineage. In this way previously exogamous units are split into intermarrying units. In all the societies mentioned, and others as well, an act of self-help that leads to negotiations if the parties belong to closely related lineages might lead to war if they are members of independent – though not necessarily geographically far apart – lineages. Such observations are indications of the flexibility of primitive social structures. They give a clue to the way in which internal adjustments are made from time to time in those structures, either in response to changing pressures from without or through the momentum of their own development. They suggest how such societies can remain stable in the long run without being rigid. But this verges on speculation.

The contributions to African ethnography mentioned in this paper are only a small and arbitrary selection from a truly vast amount of new work that is now going on in several countries. My aim has been to suggest how this work links up with a theoretical approach that is much in evidence among British social anthropologists. It is perhaps needless to add that this approach is also being actively applied by American, French, Belgian and Dutch anthropologists concerned with the problems of social organization. What I wish to convey by the example of current studies of unilineal descent group structure is that we have, in my belief, got to a point where a number of connected generalizations of wide validity can be made about this type of social group. This is an advance I associate with the structural frame of reference. I wish to suggest that this frame of reference gives us procedures of investigation and analysis by which a social system can be apprehended as a unity made of parts and processes that are linked to one another by a limited number of principles of wide validity in homogeneous and relatively stable societies. It has enabled us to set up hypotheses about the nature of these principles that have the merit of being related directly to the ethnographic material now so abundantly at hand and of being susceptible of testing by further field observation. It cannot be denied, I think, that we have here a number of positive contributions of real importance to social science.

4

Descent, Filiation and Affinity[1]

Professor Firth (1957) and Dr Leach (1957) have drawn attention to various ambiguities in current theories of kinship and descent. I am tempted to comment because Leach refers specially to a paper in which I tried to appraise the state of research and theory on the structure of unilineal descent groups in 1953 (Fortes 1953b). First, however, Leach's misinterpretations of some of the arguments put forward in my paper need to be corrected. I fail to see how my paper puts an 'exaggerated emphasis upon the principle of descent as the fundamental principle of social organ- ization in all relatively "homogeneous" societies' (Leach). On the contrary, in discussing the difficulty of defining a homogen- eous society, I point out that 'societies based on unilineal descent groups are not the best in which to see what the notion of social substitutability means'. Nor do I see how Leach reaches the conclusion that I 'disguise' ties of affinity 'under the expression "complementary filiation" '. He would, I hope, agree that there is a fundamental difference, indeed antithesis, between a tie of marriage and a tie of filiation. If this were not the case, the incest taboo would be nonsensical. Does he really believe that a person 'is related to kinsmen of his two parents' ... 'because his parents were married' and not 'because he is the descendant [I would substitute "child"] of both parents'? In many societies, as is well known, a person's jural status as his pater's son (daughter) derives from the fact that his pater was his mother's legal husband at the time of his birth. But kinship terminologies and customs show incontrovertibly that if he is entitled to claim kinship with his pater's kin it is by virtue of being his pater's jurally recognized child, *e.g.* on account of bride price. The *marriage* of the parents

[1] Reprinted from *Man*, 1959, 309 and 331. This paper was completed during my tenure of a Fellowship at the Center for Advanced Study in the Behavioral Sciences, Stanford, California. I am indebted to seminar discussions at the Center for stimulus in clarifying the argument, to Professor Raymond Firth for valuable criticism, and to Dr Leach for patient comment on an earlier draft.

is sometimes not even essential for the recognition of paternity by the genitor and the consequent acknowledgement of kinship between the 'illegitimate' offspring and the kin of the genitor. The Ashanti, like many peoples with matrilineal descent systems, and the Lozi, like others with 'cognatic' systems, are cases in point (Fortes 1950; Gluckman 1950).

Leach asserts that 'for Fortes, marriage ties as such do not form part of the structural system'. My paper is quite explicitly and primarily concerned with the *internal* structure of unilineal descent groups, not with the typology of total social systems. The generalizations which I tentatively put forward were relevant to this context. I made no particular claims about their bearing on problems of marriage and divorce. I do not, therefore, see what leads him to his conclusion. It turns, I suspect, on an implied difference between us in what we understand by 'structural system'. Anyhow, he has clearly misinterpreted what I said about the connection between marriage and what I called 'a complex scheme of individuation' among the Tiv. The sentence immediately preceding the one that he cites makes it perfectly plain that I relate this specifically to ties of *matrilateral filiation*, not to ties of marriage. These ties arise through the marriage, or more generally the conjugal unions, of parents, but it is not marriage ties as such that distinguish sibling groups and persons within a patrilineal lineage. It is quite precisely their matrilateral kinship ties, as I demonstrated at length in my studies of the Tallensi. In referring to the Tiv, I also emphasized that they recognize several degrees of matrifiliation. I am struck by the close parallel between this observation and Leach's data on the Lakher, as represented in his fig. 2, *loc. cit.* It is satisfactory that Dr Leach concedes that the generalization implied in my statement is true, even if 'inadequate' for the data presented in his paper. He tells us that he suspects ('as I suspect myself', *sic*) that we '... must take cognizance of the political and economic context before we can give a label to a structural type'. I ventured to go further, in the paper referred to, and asserted that '... it is the approach from the angle of political organization to what are traditionally thought of as kinship groups and institutions that has been specially fruitful', and I drew particular attention to the '... connection between lineage structure and the ownership of the most valued productive property of the society'.

I do not want to make too much of Dr Leach's misreading of my paper; for, in my view, there is no fundamental divergence between us. This would have been more apparent if he had cared to take into consideration some of the publications cited in the bibliography attached to my article. His analysis of Kachin and Lakher marriage and affinal relations in terms of the 'transfer of jural control' over women and the relative 'strength' of the 'sibling link' and the 'marriage link' seems to me no different in method and principle from the analysis of Ashanti kinship and marriage previously sketched by me (Fortes 1950, 1953a). For example, I point out that 'the essential nature of the sibling tie lies, to most Ashanti, in its antithesis to conjugal ties' and that 'divorce ... is said to be due very often to the conflict between loyalties towards spouse and towards sibling' (1950). Earlier (1949a) in discussing some aspects of marriage among the Tallensi, I gave special attention to the 'Opposition between Kinship Ties and the Marital ties of spouses, and its resolution' (*op. cit.*, pp. 90 ff.). I there exemplified the proposition stated in my paper of 1953(b) that 'a common rule of social structure reflected in avoidance customs is that these two statuses [of spouse and parent in the conjugal family *versus* child and sibling in the natal family] must not be confounded'. This is a way of achieving the 'institutional accommodation of the possibly conflicting claims and loyalties as between a woman's husband and her brothers and sisters' (Radcliffe-Brown 1950). And it seems to me that the data presented by Leach fit the proposition very well.

I would maintain, therefore, that Dr Leach and I have been using the same conceptual framework. His further refinement of it in the paper under discussion does not, I believe, amount to a radical divergence between us. I expect that he would agree that we owe the 'model' originally to Malinowski's analysis of Trobriand marriage and family relations, and its validation to such penetrating studies as those of Fortune (1932), where the emphasis is on the conflict between 'brother-sister solidarity' and the strength of the 'marital grouping', and Richards (1934) where the power of 'father-right' in a family and kinship system based on matrilineal descent, is specially documented. This was, however, only a descriptive model. The analytical principles, which both Leach and I (in common with others) use, come, as I emphasized in my paper, from Radcliffe-Brown, notably his famous paper on patrilineal and matrilineal succession (1935a).

I agree with Leach that 'it is the whole nature of the concept of "descent" which is at issue', and in my view we can clarify this concept fruitfully if the distinction which I made between 'descent' and 'filiation' (1953b) is pursued further. To do this, however, we need a more precise analytical procedure than has been the rule in kinship studies.

In my paper I drew attention to the distinction between the *external* aspect of lineage structure and the *internal* aspect. This distinction, I believe, sums up a general rule of analytical procedure which can be helpful in the task suggested by Leach. Institutions, like structural units, can be viewed in terms of their internal aspect, as components of an internal system or field of social structure, specified for a particular purpose, or they can be considered in terms of an external aspect, as components of an external system of social structure. An internal system, whether it be a unit of social structure like a lineage, or an institution like marriage, or a body of beliefs, is a combination of elements or arrangement of parts that operates as a unit in external relations. The analogy with the organs of the body is obvious.

In the study of kinship and descent we are dealing with institutions that operate in various fields, or, as I now prefer to say, domains, of social structure (Fortes 1958). In their external aspect it is their function in the domain of politico-jural structure that is of major importance; in their internal aspect it is their place in the domain of domestic relations that is crucial. In previous publications (1951, 1954) Leach investigated Kachin marriage and affinity as institutions in the politico-jural domain. He showed how these *descriptively* speaking 'kinship' institutions are utilized to maintain a fixed hierarchy of political rank and power between the genealogically and politically independent patrilineages of a Kachin community. Here marriage is an enjoined transaction between corporate units bound to one another as if by a treaty of alliance. It is a transaction in the politico-jural domain and in this context the internal composition of the units involved is a subsidiary issue. It is easy to visualize identical politico-jural marriage transactions taking place between units not *internally* organized by a descent criterion. As Leach puts it, the 'mutual status relations' of lineages *in the political system* are defined through these marriage rights and obligations, which are enforce-able and exclusive. A significant index of this is that in these external relations of lineages the spouses represent their natal lineages, as corporate units.

An analogous structural arrangement is followed by the Tallensi (Fortes 1940, 1945); but among them the ties claimed between corporate lineages are conceptualized as perpetual *kinship bonds* originating in matrilateral filiation and perpetuated by the lineage principle. The Kachin pattern has closer parallels in African societies where there are aristocratic lineages, clans or classes which receive periodical tribute or services from client popu-lations in return for protection and the maintenance of political and ritual security. An instructive instance is provided by the Banyankole (Oberg 1940). Among them marriage was tradi-tionally *prohibited* between members of the superior Hima pastoralist caste and the inferior Iru agricultural caste, lest the links so created should undermine the supremacy of the former. But this prohibition, says Oberg, was deemed to play a part in 'keeping the relationship alive' between the two castes. Thus a marriage rule that is the opposite of the Kachin rule appears to serve the same end of maintaining mutual status relationships between differentiated structural units in the politico-jural domain.

Why should marriage between members of a superior caste and members of an inferior caste be regarded as a potential threat to the supremacy of the former? Primarily because title to membership of the superior caste is derived from 'bilateral' filiation. Both of his parents must be members of the caste by birth for a person to be a member by right of birth. The threat lies in the jural value given to matrilateral filiation. Even if it does not confer status in the politico-jural domain, as is the case in a segmentary lineage system like that of Tallensi (*cf.* Fortes 1945) it carries residual rights with their concomitant moral and ritual claims and associated affective bonds. These belong primarily to the domestic domain, or what is sometimes spoken of as the sphere of interpersonal or cognatic kinship (*cf.* Fortes 1949). But they derive from the jural rule that a woman never wholly forfeits her status as a member by right of birth of her natal lineage. Taking the politico-jural domain as a whole, the rami-fications of matrilateral kinship ties result in what I previously called 'the web of kinship'; and as I showed for the Tallensi, the web of kinship plays a critical part in restricting warfare between corporate lineages and thus helps to maintain the balance of political power in the total system. The point at issue here is the result, not of marriage, but of ties of cognatic kinship mediated by combinations of links of matrilateral filiation operating in a complementary way to patrilateral filiation and patrilineal descent.

How this would work out among the Banyankole is neatly illustrated in a case record for which I am indebted to Dr D. J. Stenning, who recently completed a tour of field work among them. Briefly, *A* is a highly respected senior functionary in the Ankole government. A well-to-do cattle-owner, he lives like a Hima, observes Hima custom in his domestic arrangements and public activities, and is reputed to be a Hima. But there are signs that he is not accepted by the aristocratic Hima as one of themselves without some reservations. In private he explains that he is a member of both a Hima subclan and an Iru subclan, which is of course impossible, *de jure*. It is due to the fact that he is originally Iru by patrilineal descent but Hima by maternal connection. *A*'s grandfather was an Iru hunter who performed valuable war services for the ruling Mugabe. The King rewarded him with a gift of cattle, the hallmark of the Hima caste, and in addition gave him a Hima girl as wife. Significantly, the girl was an orphan of probably low rank, without close kin to be concerned with her marriage. The hunter's half-Hima son repeated his father's history and this continued in the next generation. *A* is therefore three-quarters Hima by successive steps of matrilateral filiation, but one-quarter Iru by patrilineal origin. Strictly speaking he belongs to his Iru ancestor's Iru subclan but he is accepted as a member, with limited rights, particularly in ritual matters, of his mother's Hima subclan. This, he adds wryly, is due to the influential office he has achieved in the political sphere. *A*'s case is not unique. But a part-Hima by matrilateral kinship cannot normally acquire Hima status, which presupposes membership of a Hima patrilineal subclan. Analogously, Iru who keep cattle are becoming numerous nowadays, but they do not thus win recognition as Hima. Hence it is not surprising to find that there is an interstitial element of Hima-Iru. This, following the basic caste division, is a category in the internal domain of caste. *A*'s story shows however that caste status does not exclusively determine status in the politico-jural domain. But as political office is ideally the preserve of the Hima, his status is legitimized, and consistency is maintained, by recognizing his matrilateral ties as effective for his caste membership. From his point of view, matrilateral filiation – not, be it noted, his or any of his predecessors' marriage – permits him to become attached to his maternal Hima subclan for political status purposes.

This example shows how essential it is to specify the structural domain with which we are dealing when we examine kinship and descent institutions. This is the only way to avoid the kind of ambiguities which, in my judgement, mar the effect of Leach's otherwise penetrating application of the Radcliffe-Brown 'model' of marriage relations in the paper under discussion.

Thus a critical variable in Leach's analysis is the balance established between the sibling link and the marriage tie. How must we interpret the proposition that 'a sibling link is "intrinsically" more durable than a marriage tie' and the comparisons made between them as to relative 'effectiveness', 'strength' and 'fragility'?

Since Leach makes no reference to Radcliffe-Brown's classical formulation of the hypothesis (*loc. cit.*, 1935; see also *ibid.*, 1950, pp. 77ff.) it is perhaps worth quoting *verbatim*:

> Thus the system of patrilineal or matrilineal succession centres largely around the system of marriage. In an extreme matrilineal society a man has no rights *in rem* over his children, though he does usually have certain rights *in personam*. The rights remain with the mother and her relatives. The result is to emphasize and maintain a close bond between brother and sister at the expense of the bond between husband and wife. Consequently the rights of the husband over his wife are limited. In an extreme patrilineal society we have exactly the opposite. Rights *in rem* over the children are exclusively exercised by the father and his relatives. The bond between husband and wife is strengthened at the expense of the bond between brother and sister. The rights of the husband over the wife are considerable; she is in *manu*, under his *potestas*.

It is true that Leach's discussion of the Lakher situation can be read as an exemplification of Radcliffe-Brown's formulation. But the general drift of his analysis is to put the emphasis on the sibling bond itself as the irreducible factor; whereas Radcliffe-Brown derives the relative 'strengths' of the sibling bond and the marriage bond from the descent system. In other words, whereas Leach considers the sibling bond as founded in the domestic domain, with the implication that its 'strength' or 'fragility' is an intrinsic characteristic, Radcliffe-Brown regards it as primarily an institution in the politico-jural domain, and therefore varying with the nature of the descent principle. It is clear that Radcliffe-Brown is talking about the relative *jural* 'strength' of the sibling bond, as the focus of descent-group rights, and the marriage bond as the focus of paternal rights. Leach, on the other hand, seems to base his argument on other criteria. 'Durability' may mean priority in the life cycle, since one is a sibling before one is a spouse; and 'strength' could be interpreted as a measure of the affective bonds generated in the fraternal and filial relationships. I fell into the same ambiguities in my analysis of the status of women among the Tallensi. I demonstrated that a woman never forfeits her status as a member of her natal family and lineage (1945, ch. IX) and that the bride-price is an instrument for separating her role as wife from her role as daughter (1949, p. 92). But, I did not elucidate in what respects she retained filial rights and obligations in her lineage after marriage and how these affected her conjugal relationships. The point is clearer if we remember that in some patrilineal descent systems (*e.g.* among the Zulu, Gluckman 1950, and the Chinese, Yang 1945) the affective bonds of brother and sister can remain undiminished after a sister's marriage though her jural bonds with her natal kin are all but completed severed.

I would argue, therefore, that the differences between the relative 'efficacy' of sibling ties, marriage ties and affinal ties, as between Lakhers and Kachins, that are postulated by Leach, must derive from differences in their systems of descent and kinship. For what is at issue is relative jural efficacy, not the affective bonds; and the jural significance of any relation of kinship emanates, I maintain, from the politico-jural domain.

In this connection Leach claims that 'by ordinary criteria the Lakher seem to be just as patrilineal as the Kachin'. But he does not state his criteria, or give us the relevant data. What, for example, is the generation span of the lineage in each of the two groups? Are succession and inheritance rights identical? And what are we to make of the fact that Kachin patrilineal clans and lineages are strictly exogamous (Leach 1954, pp. 73f.) whereas among the Lakher, there is reported to be no bar to people of the same clan marrying (Parry 1932, p. 293)? If, as Leach says, they are unlike those of the other two peoples in not being 'ordinarily of the segmentary type' this surely indicates a critical difference in 'patrilineal ideology', epecially in relation to rights over persons. In a segmentary lineage system rights over persons are correspondingly segmented and subject to a hierarchy of jural control. They are not so distributed in an unsegmented system of shallow generation depth. This has a direct bearing on the mode of distribution and transfer of marriage rights. In an unsegmented lineage jural authority to receive the critical marriage payment is likely to be held by the lineage head; in a segmented lineage this right may be vested in a woman's father, subject to consent of a superior lineage authority.

This brings up the question of 'father-right'. If I understand him aright, Leach agrees with Gluckman that 'father-right' can have different 'degrees'. Surely father-right can be defined precisely as the rights over his legitimate child, or any person in that relationship to him, vested in a man in his capacity as *pater*. The rights which a man has over his wife are derived from the rights which her father had over her as daughter. For it is always a father (or his proxy or matrilineal equivalent) who permits marital rights to arise and transfers them. If a woman's brother does this, it is never *in propria persona* but as the representative of the original holder of father-right. But the notion of degrees of father-right can be misleading. For it is not a matter of amount of authority but of different combinations of elements of right, as I think is apparent from Leach's remarks. The question that really matters is what is the source of father-right? Matrilineal systems give us the answer very clearly. Father-right is a function of the descent system. It emanates from the politico-jural domain. For a matrilineal father often has domestic authority over his wife and specific moral and ritual authority over his children. This arises from his role as provider for the family and as begetter and upbringer of his children. But they are not his descendants for politico-jural purposes of inheritance, succession and citizenship, and therefore do not come under his jural authority (*cf.* Richards 1950). A matrilineal father's rights over his children are based on the principle of filiation, the mother's brother's on the principle of descent. If a Lakher father has 'less marked father-right' than Ordinary Jinghpaw father, it could well be argued that the former's paternal status has elements of matrilineal fatherhood in it if the latter's status is that of a 'normal' patrilineal father.

We do not, however, have to resort to matriliny to explain this. Leach's data, to my mind, show that the difference is related to the more extensive recognition of complementary filiation in the Lakher than in the Jinghpaw patrilineal system, and this, I suggest, is due to the greater jural 'efficacy' of a woman's filial status in her patrilineage among the former. The 'strength' of the sibling tie is a *result* not the *source* of the filial status which is established by the descent system and is expressed in the distribution of jural control over a woman's offspring between her husband's and her own lineage. Leach incidentally, seems to contradict himself on this point. First he rejects an interpretation in terms of 'a "submerged" principle of matrilineal descent', but later, questioning the adequacy of an analysis in terms of my contrast between filiation and descent, he concludes: 'In our usual terminology they (*i.e.* Kachin and Lakher) are patrilineal systems in which the complementary matrilineal descent line assumes very great importance.'

The argument of the preceding part of this paper leads us back to what is essentially at issue between Leach and Gluckman. It is, to paraphrase Leach, the whole nature of the institution of descent. In my view the ambiguities that surround the concept of descent, as we have been accustomed to use it, arise from confusing two analytically distinct institutions, that of descent in the strict sense and that which I have called filiation. As I pointed out in my paper of 1953(b), 'filiation – by contrast with descent – is universally bilateral'. Before that, in discussing the connection between descent and parenthood in Tale lineage structure (1949, chapter V) I distinguished two aspects in the continuity of the lineage through time. One I described as 'the straightforward, cumulative continuity of descent, epitomized in the notion of patrilineal descent', the other as 'the dialectical continuity of the filial generation ousting and replacing the paternal generation'. I continued: 'In terms of patrilineal descent father and son are identified with each other and united by common interests; in terms of the sequence of generations, they are not' (*loc. cit.*, p. 135).

This descriptive distinction can now be refined in the light of subsequent research, especially such studies as those of Pehrson (1957), Freeman (1958) and Dunning (1959) on what are commonly called bilateral or ambilateral kinship systems.

By filiation, in its primary sense, I mean the 'fact of being the child of a specified parent' as the *Oxford English Dictionary* puts it (quoted by Freeman, *loc. cit.*, 1958). More precisely, it denotes the relationship created by the fact of being the legitimate child of one's parents. Since the great majority of societies[1] give jural recognition to the parenthood of both parents, filiation is normally bilateral, or as we might even say, equilateral. As a relationship in the internal domestic domain it denotes the specific moral and affective ties and cleavages between successive generations (due allowance being made for sex differences) that arise from the facts of begetting, bearing and, above all, exercising responsibility in rearing a child. But it is also legitimized by sanctions derived from the politico-jural domain. These define a child as the eventual social no less than the biological replacement of both his parents. They ensure that he succeeds, by right of filiation, to those components of their jural status that are valid in the domestic domain. Thus a matrilineal son does not succeed to those components of his father's status that are valid in the politico-jural domain, for example, his status as a potential office-holder in his clan, or his status as mother's brother to his sisters' children. But he may, as in Ashanti, be taught his craft by his father and succeed him in that capacity. Thus filiation is essentially the bond between successive generations – a bond, as we well know, compounded of rights and identifications epitomized in rules of inheritance and succession, on the one hand, and of the cleavages symbolized by the incest taboo and in customs of respect and avoidance. It is, obviously, a critical factor in the definition and internal relationships of the sibling group. Persons are siblings in the domestic domain by virtue of common filiation and with polygynous marriage they are usually graded according to whether their common filiation is unilateral or bilateral. Clearly, also, what cognisance is accorded to filiation in the politico-jural domain depends on the nexus that links that domain with the domestic domain.

[1] I should prefer to say 'all societies' but use the more guarded formula to cover cases like that of the traditional Nayar family system for which it might be argued that filiation to a particular pater is not jurally recognized.

I turn to consider descent as a structural principle. Descriptively, descent can be defined as a genealogical connection recognized between a person and any of his ancestors or ancestresses. It is established by tracing a pedigree. A person's entire pedigree, if it is known, of necessity ramifies bilaterally with each successive generation of ancestry included, the total number of ancestors, male and female, being given by the well-known formula 2^n where n is the number of antecedent generations reckoned from and including the parents. But as is well known pedigrees are as a rule only selectively utilized for defining and identifying persons in any society. Whatever the rules of selection and limitation may be, only full siblings can have a common pedigree. But any two or more persons whose pedigrees converge in a single common ancestor can be said to be linked by descent.

It is of course not essential to establish pedigree in order to claim or even demonstrate descent from an ancestor. At one time in their history the whole population of Pitcairn Island could presumably claim common descent from a known group of ancestresses. They would not have needed to demonstrate this by reckoning pedigrees. They would only have had to show that there had been no immigration since the arrival of the founding ancestresses and their husbands and that sexual relations were freely permitted except within the sibling group. Thus isolation or even enclavement, as of a religious minority like the Jews in an East European ghetto, or a caste, as long as it is coupled with obligatory endogamy, can produce a common descent group none of the members of which need to know or establish their pedigrees in order to identify themselves or be accepted as members. Such a group would, of course, be a bilateral descent group.

Looked at from the inside, that is from ego's position, a pedigree is established by counting successive steps of filiation. We may call it a unit of serial filiation, and it requires a minimum of two successive steps of filiation to ensue. Looked at from the outside, from the point of view of the significance attached to any particular form of pedigree in a given society, or, to put it in another way, from the point of view of the social relations between persons that are governed by the recognition of pedigrees, a pedigree is the charter of its bearer's descent. A descent rule states which of the two elementary forms of filiation and what serial combination of forms of filiation shall be utilized in establishing pedigrees recognized for social purposes. Thus a rule of patrilineal descent states that only pedigrees made up exclusively of successive steps of patrifiliation are recognized as conferring descent for the particular social purposes in question.

In short, whereas filiation is the relation that exists between a person and his parents only, descent refers to a relation mediated by a parent between himself and an ancestor, defined as any genealogical predecessor of the grand-parental or earlier genera-tion. A grandparent is therefore a person's closest ancestor; and this, as we know, is often shown in kinship terminologies, as among the Ashanti and the Tallensi (*cf.* Fortes 1950, 1949a)

Rules of descent, as opposed to the rule of filiation, in my judgment, always emanate from the politico-jural domain, or in descriptive terms, from the total social structure. Descent is, as I have elsewhere maintained (1953b, 1955), fundamentally a jural institution. The distinction is often apparent in the difference between relations with a grandparent based on the criterion of successive filiation *per se*, and relations based on that of descent, though the two statuses coincide in the same person. Tallensi, for example, joke with a parent of either parent, when the operative relationship is that of parent's parent to child's child. They are subject to a father's father's jural authority – which emphatically rules out joking – when the operative relationship is common descent in the lineage segment of which the father's father is head. The joking relationship falls within the internal domestic domain; the authority relationship in the external domain of the segment's jural and ritual position in relation to the other segments of the maximal lineage. Thus where descent has structural significance (as it has not, for instance, in our social structure) one of its main functions is to bind the domestic domain to the politico-jural domain, as I pointed out in my paper of 1953(b). This is obvious from the fact that certain critical attributes of the jural status of members of domestic groups (*e.g.* the rights and capacities of paternity which entitle a father to administer corporal punishment to his minor offspring and oblige him to support them) are legitimized by politico-jural rather than purely moral, ritual or affective sanctions.

I might add, incidentally, that this analytical distinction has a bearing on the theory of lineage systems which I had not pre-viously perceived. In the *Dynamics of Clanship* (ch. III) I defined the minimal lineage as comprising the children of one father. This definition has not, as far as I know, been generally accepted. Of course from the internal point of view, in terms of their domestic relations, the children of one father in a patrilineal descent system constitute the nuclear sibling group. This is a definition by a criterion of filiation. But by the rule of descent, in a segmentary patrilineal system, a man's children are the minimal element in the total system, taking their politico-jural status from their whole patrilineal pedigree and from the position which this allocates to them in the total lineage system. In such a system the internal differentiation between siblings by patrifiliation originates in the domestic domain but also receives jural sanction.

In view of the intrinsic connection between kinship institutions and political structure, it is not surprising that a good test of the distinction which I am making lies in the rules governing citizenship in a political community. In European societies neither filiation nor descent necessarily affects the issue. Citizenship is acquired by birth in the territory of the State or by naturalization (*i.e.* civic adoption). The citizenship of one's parents or ancestors is then irrelevant.[1] But in the Athens of Pericles the rule was very different. It is recorded that Pericles made a law in 451–50 B.C. restricting the franchise of Athens to persons who could prove Athenian parentage on both sides; and the *Cambridge Ancient History* (vol. v, pp. 102, 168) describes this as an undemocratic and reactionary step since it went back on the reforms of Cleisthenes which had admitted both resident aliens and the offspring of mixed marriages to citizenship. Thus the Periclean reform introduced citizenship on the basis of bilateral filiation in the place of citizenship irrespective of genealogical connection. Now take the case of an Ashanti born outside his own country. He has title to citizenship in the chiefdom where his matrilineal lineage is domiciled. He asserts this title by demonstrating first that he is his mother's child, then that his mother is or was a member of the matrilineage of which he claims membership. This is citizenship by descent and with it go rights of inheritance and succession

---

[1] The position of children born to British parents in a foreign country forms a special case. They have dual citizenship, being entitled to claim British citizenship by filiation, but obliged to acknowledge citizenship of their country of birth when in that country.

(*cf.* Fortes 1950). He cannot claim citizenship in his father's community by reason of being his father's son, for that does not make him a member of his father's matrilineal descent group, but he can, on this basis, claim gifts given to him by his father, residence in his father's house and chiefdom, and protection by his father's kin. These rights come through patrifiliation. The sanction for this is the special ritual and moral identification with his father which identifies him also with all the other descendants, male and female, by *successive steps of patrifiliation* of his father's father's father (*cf.* Fortes 1950). Whereas matrifiliation follows automatically from the fact of birth and automatically *creates title to lineage membership by descent*, patrifiliation ensues only from acknowleged paternity. Here complementary filiation does not establish citizen rights in the juridical sense. But it is of peculiar significance in defining capacity for unencumbered citizenship in that it is a prerequisite for normality of status as a free citizen, in contrast to the status of those who have slave ancestry. A person may be a legitimate member of his matrilineage and by that token a free citizen but if his paternity is unknown or un-acknowledged he is nevertheless socially defective through lack of the patrilateral side of the normal network of kinship ties and is in consequence ritually defective and jurally incomplete. This is a distinct handicap, if not an insuperable obstacle, to eligibility for lineage and political office. Indeed so important is patrifiliation in Ashanti that it can serve, in special circumstances, as the basis for acquiring *quasi-* citizenship in one's father's chiefdom. A man can do this by voluntarily accepting such distinctive *political* obligations as paying his share of a money levy imposed by the chiefdom and serving in its fighting forces in defence of its land and people. He thus identifies himself with his father's status in the chiefdom as if he were his father's representative. But he is not eligible on this account for any office or rank that may be vested in his father's lineage even if it had been held by his father. Nor can he transmit his acquired *quasi*-citizenship to his offspring by right of filiation.

To sum up, filiation originates in the domestic domain, descent in the politico-jural domain, but filiation may confer title to status (which means rights and capacities) in the politico-jural domain. What is thus conferred is entitlement to activate those elements of politico-jural status carried with him or her into the domestic domain by the parent in question. Where complementary filiation is recognized this rule can be extended beyond the first degree of filiation to include successive steps of filiation documented by a pedigree. Since descent confers attributes of status relating to a person's place in the external social structure it is bound to operate by placing persons in categories or groups. Descent groups exist to unite persons for common social purposes and interests by identifying them exclusively and unequivocally with one another. Descent operates where the total body of rights and duties, capacities and claims, through which a society achieves its ends, is distributed among segments, or classes, which are required to remain relatively fixed over a stretch of time in order that the social system shall be able to maintain itself. Empirically, descent groups are constituted by the fact that all the members of a group in a given society have the same form of pedigree and all their pedigrees converge in a single common ancestor or group of ancestors. Theoretically, they are necessarily corporate groups, even if the corporate possession is as immaterial as an exclusive common name or an exclusive cult.

It is obvious that in systems where a sibling succeeds or inherits 'in preference to', *i.e.* by priority of right over, a child, descent is the critical factor; for a sibling is closer to the source of the deceased's 'estate' – a common ancestor – than is a son or daughter. But where succession and inheritance devolve on sons or daughters 'in preference to' siblings, this is governed by the rule of filiation. The rule of so-called primogeniture is in fact, analytically speaking, a rule of succession by filiation. The Tswana illustrate this excellently. A chief's legitimate successors are, first, his sons in order of birth and failing them his brother, the implicit logic being that the step of filiation which made him heir is extinguished by his failure to perpetuate himself in sons. The chiefship is then treated as if it had reverted to the previous holder by right of filiation and devolves from him on *his* oldest surviving son (*cf.* Schapera 1937, pp. 53 ff., 177, 191, etc., for the data; the interpretation is mine, not Schapera's). Descent, reduced to its elementary significance as the unequivocal source of title to class membership among the Tswana nobility, enters as a factor of opposition to filiation. It establishes what might be called a right to a place in the queue of potential successors, this right being virtually the basic corporate possession of the noble class. It serves as the justification for reserved claims on the chiefship.

From what I have said it can be seen that the classical mother's brother – sister's son relationship, in societies with patrilineal lineage systems, flows from the recognition of matrifiliation as the complementary structural factor to patrifiliation. The sister's son's *quasi*-filial status among the Tallensi, for example (*cf.* Fortes 1949, p. 30 *et passim*), arises from the fact that his mother carries with her into marriage and parenthood her filial (and *eo ipso* sororal) status in her natal minimal lineage *but not her descent status*, vestigial though that is by reason of her sex. It is not a question of affective bonds that *grow up out of* frequent contacts between sister's son and mother's brother (as *e.g.* Homans and Schneider (1955) seem to think) but of *permitted*, indeed *enjoined* sentiment, that canalizes into action the claims and privileges, material and ceremonial, vested in and transmitted through 'daughterhood' and 'sisterhood'. We might almost call them *ex gratia* claims and privileges. A sister's son may be deeply devoted to his mother's brother but he does not get from his mother her obligations, *acquired by descent*, to observe the particular exogamic prohibitions and totemic taboos of her lineage, nor does he get from her the right to take the same part in sacrifices to her lineage ancestors as she has. As the Tallensi themselves put it, patrilineal descent does not pass through a woman, hence property and office cannot pass to her children. But filiation, her status in the domestic domain, does pass to her children, and this is the basis of their rights – enforced by moral sanctions – to gifts from their mother's brother and to such customary acknowledgements of a special status as shares in the meat of sacrifices offered by their mother's patrilineal kin.

This analysis applies also, *mutatis mutandis*, to the matrilineal case. We can see how vital it is to determine the jural attributes of 'daughterhood' and 'sisterhood' before we make assumptions about the 'intrinsic' priority of the brother-sister bond over the conjugal bond. In the limiting case at the patrilineal pole, marriage may entail almost complete severance of a woman from her natal family and the virtual extinction of her 'sisterhood' and 'daughterhood' in any jural sense – though not her affective attachments to her parents and siblings. This appears to happen in some parts of rural China, where a woman's jural status and rights in her natal family are insignificant but she loses even these when she marries.[1] There appears to be no institutionalized recognition of matrifiliation, divorce is 'almost impossible' (Yang, *op. cit.*, p. 240), and patrifiliation is the basis of an elaborate patrilineal lineage and clan system. The limiting case at the matrilineal pole, as with the Nayar (Gough 1952), gives an exact opposite scheme of relationships.

With both patrilineal and matrilineal descent systems the crucial test, at any rate in Africa, is the status of the slave spouse, as I note in my paper of 1953b. A slave is jurally defined as both kinless and a non-citizen. Hence a slave spouse is wholly and unreservedly under the jural authority of his or her spouse or spouse's jural superiors. This ensues in the paradox of pure and unlimited 'father-right' in matrilineal systems, as for example among the Yao (Mitchell 1956, p. 69), and indicates how the degree of 'father-right' (or 'avuncular right') is determined by the kind and degree of complementary filiation. For a slave spouse can create no connections, either by descent or filiation, for his or her children, and both the slave and his or her children are therefore completely subject to the husband/father's (or, for a male slave, wife's brother's or uncle's) authority. Yet the purely affective and moral elements of filiation, engendered by the reciprocities of parental care and filial dependence, are as patent as, and probably even stronger than, happens in the case of both parents being free. Thus since a slave is by jural definition nobody's son or daughter, brother or sister, his or her children, though free by descent through the free parent, have none of the attributes of status derived from complementary filiation. Hence they are always incomplete jural personalities, liable to discrimination and disabilities. It is significant that these conditions may be exploited by chiefs to bind their slave children unreservedly to themselves and so to maintain power which the descent and kinship system as a whole normally keeps in check – again through the mechanism of complementary filiation (*cf.* Mitchell, *op. cit.*).

---

[1] 'A girl has no status whatsoever in the family of her own parents. Her father and mother and brothers may love her very much. It is recognized that she is not a permanent member of the family and can add nothing to the family fortunes. She is destined to become a wife and daughter-in-law in another family for whom she will work and bear children' (Yang 1945, p. 104). Freedman (1958, pp. 30 f.) comes to broadly the same conclusion after a careful assessment of the best available evidence.

Finally and most pertinently, divorce is of course both *de jure* and *de facto* impossible for a slave spouse.

These two examples bring out another important point, on which, I believe, Leach and I are really in full accord. Complementary filiation is a function of affinal relations, or rather, both are functions of the distribution of rights, duties and sentiments among the persons and kin units joined by marriage. Without marriage (as in concubinage) there are no affinal relations; and where there are no affinal relations as in the case of a slave spouse, there can be no complementary filiation for the offspring of the marriage. Complementary filiation can be thought of as the kinship reciprocal of affinal relationship in the marriage tie; and this is perhaps what Leach has in mind when he says that I disguise the latter under the former.

Though there is no divergence in principle between Dr Leach's views and mine, there is perhaps one point of difference in emphasis. Leach thinks that it is the relationship of marriage and its concomitant relationships of affinity that form the 'crucial' link between corporate descent groups in the Kachin-type system. I would put it the other way round and say that marriage and affinity are the *media through which* structurally prior politico-jural alliances and associations are expressed and affirmed, and I would contend that they are effective as such media because they give rise to matrilateral kinship bonds. This is an argument from first principles, not from the data presented by Leach. The principle is that kinship, being an irreducible factor in social structure, has an axiomatic validity as a sanction of amity and solidarity. It seems to me that a social structure based on an association of exogamous, corporate patrilineal descent groups is not likely to hold together in a permanent political system unless it is either subject to some form of overriding, centralized government, or is knit together in the field of dyadic social relations by a web of kinship ties that counterbalance the centrifugal tendencies of the descent groups; and the main mechanism involved is complementary filiation, not marriage as such.

As African segmentary systems illustrate so clearly, marriages commonly take place between members of politically autonomous and mutually opposed, if not hostile, descent groups. Affinal relations are perfectly compatible with such a state of political relations. They do not, of themselves, constitute either a moral or a political bond between groups. It is the kinship ties generated through marriage that constitute such bonds; and we can see confirmation for this view in the fact that marriage right is subsumed under kinship (cross-cousin) right in Kachin-type marriage. Thus marriage is a means of implementing a relationship of amity derived from real or putative matrifiliation.

As we have seen, in a segmentary patrilineal descent system matrifiliation endows a child with attributes of jural status subsuming claims on and obligations to its mother's lineage, without which it cannot be a normal jural person. It seems that this is more elaborately institutionalized in Lakher than in Ordinary Jinghpaw social structure. I suggest that this is the reason why the sibling link may well be jurally – and, in consequence, affectively – so 'strong'. The 'lien' which every person has on his mother's and mother's mother's natal lineages derives from the inextinguishability of his mother's and mother's mother's filial and sororal status in their natal lineages; and it is essential for a person, whether man or woman, to be able to exercise the normal rights and capacities of his status. This implies that the jural authority and responsibility vested in a father and his jural superiors is not unrestricted and undistributed but is shared with the mother's brother and his lineage superiors in the manner previously indicated.

Hence the diminished 'father-right' noted by Dr Leach. With the Ordinary Jinghpaw matrifiliation seems much more restricted in significance. This would be consistent with a social structure in which distinctions of caste, class or rank are strictly aligned with exogamous descent groups. Then the only basis for membership of a caste, class or rank would be patrilineal descent. If matrifiliation is minimized to such an extent that a person has only the minimal moral or ritual connection with his mother's lineage, there is no loophole by which he can claim admission to any form of membership in it and therefore in its caste, class or rank position. We have seen that among the Banyankole successive matrifiliation with the Hima caste enables a man of Iru patrilineal origin to achieve *quasi*-membership of a Hima sub-clan and thus to 'rise' in rank. On this hypothesis it should be easier and not uncommon for an individual (or sibling group) to claim some form of membership in a lineage with which he is connected by matrifiliation and so to change his 'class' rank.

The Lakher data could be interpreted on these lines, but I find Leach's analysis confusing, owing, perhaps, to his use of the concepts of 'class' and 'status' interchangeably. The purpose of 'class hypogamy' is said to be 'raising children of relatively high status'. If, as Leach implies, the Lakher patrilineage is exogamous and is also a unit of 'class' (*sc.* 'status') rank, and a man's sons by 'marriage proper' (*i.e.* hypogamous marriage) have 'higher status' than he has, they cannot belong to his 'lower' ranking patrilineage. For it does seem that 'class' (*sc.* 'status') rank is derived by matri-filiation through the mother, not by descent from the father, or else the purpose attributed to hypogamous marriage could not be fulfilled. On the other hand, if a man's 'higher status' children do not have some kind of membership claim in their mother's lineage by right of matrifiliation but are strictly members of their paternal lineage, then each lineage must be internally stratified by 'class' (*sc.* 'status') and it is not a unit of 'class' rank.

I mention these difficulties not to quibble over Leach's analysis but in order to point out that they can be resolved if the distinction between the internal domain and the external domain of the lineage is borne in mind. In the Tale-type patrilineal descent system, the internal differentiation of the lineage is effected by segmentation and stratification by reference to complementary filiation and generation. The Lakher, to judge by additional data kindly provided for me by Dr Leach, carry the process further by using the criterion of matrifiliation for ranking as well as for differentiating persons within the lineage for such jural purposes as inheritance. With three degrees of matrifiliation, and two grades of legitimacy to exploit, they presumably achieve a highly refined ranking system. But this has only internal validity. It does not apparently confer rank in the external politico-jural domain. There a person's status derives strictly from his patrilineal descent, on the principle that every member of a corporate lineage is equal to every other and represents the whole lineage in relation to other autonomous lineages. This implies a stricter demarcation between the internal domain of the lineage and the politico-jural domain than is found in Ankole. And it seems to me that it could be maintained only if the filial rights transmissible by a woman to her offspring were confined to the domestic segment of her lineage and not extended to the entire lineage as a unit of political rank.

This has a bearing on the final paragraph of Leach's paper, but

I am not sure that I understand his point. Surely the 'ongoing structure' of any and every *unilineal* descent group, *looked at from within*, is determined by rules of descent. Membership of the 'ongoing' descent group is independent of the kind of marriage, whether enjoined, preferred or free. If, however, we consider unilineal descent groups from the outside – that is how they fit into a social structure which embraces a number of such groups – we are in the field of politico-jural institutions; and in this context we may need several, not just two categories.

I have not discussed Dr Leach's starting point, the problem of divorce. I would only like to suggest that complementary filiation is a relevant variable in considering both the *de jure* permissibility and the *de facto* incidence of divorce. I would suggest that divorce is correlated with the degree to which a person has jural status that is independent of his or her status as a spouse. For a woman the significant factor is the degree to which she retains her status as daughter and sister after marriage, for this determines her claims on support as well as her jural status outside the conjugal relationship. In our society this para-marital jural status derives from the laws of legal majority and citizenship. In many primitive societies it derives from the pre-marital status conferred by filiation and descent. Hence a woman who has no status by complementary filiation, even if she has a descent status, is undivorceable. This is proved by the case of the slave spouse and the Chinese case.

The foregoing discussion of filiation and descent has a bearing, I believe, on the problems raised in Professor Firth's paper (1957) and I should like to try to restate them in terms of the point of view which I have developed. It is necessary to note, in this connection, that Firth's concept of 'affiliation' does not correspond to my use of the term 'filiation'.

Let us consider the Maori *hapu* (Firth, *loc. cit.* and 1929, especially pp. 97 ff., 368 ff.). Looked at from the outside, in terms of its status in the politico-jural domain, the *hapu* is, in Professor Firth's words, an analytically irreducible segment of the total tribal structure. In this context it is a territorial and *ipso facto* political unit, structurally demarcated, as it seems, by external political cleavages expressed in opposition to like segments, often in the extreme form of warfare. From the outside, as a unit in the political domain, its internal composition is immaterial to its structural relations with other like, or subordinate or superordinate segments. It may be represented by a chief and defended by an army. How the chief is selected and the army constituted are internal matters and it would be theoretically possible for different *hapu* of the same tribe to have different institutions for these purposes. In fact, of course, all are subject to the structural norms and the values valid for the total social system, both as a whole and in its parts, and legitimized by sanctions derived from the total society. Kinship and rank are the critical values at issue and they enter into every field of social structure, though in different ways, depending on the other variables simultaneously involved.

To see how this operates let us consider the internal structure of the *hapu*. There are two questions. First, how does one become a member of a *hapu*? It seems to me that the normal source of title to membership for the free person is filiation; and with the Maori, filiation to either parent confers such title. This may be related to the relative jural equality of men and women or to the absence of corporate unilineal descent groups or both. The *hapu* itself, as Firth made abundantly clear in 1929, is not demarcated by a descent boundary as, for instance, an Ashanti or a Tale lineage is, but by a political and territorial boundary. Within the tribe, therefore, persons can enter a *hapu* from the outside (*i.e.* from another *hapu*) if they have the jural status to cross the boundary; and this can be claimed by filiation if either parent has title to citizenship in the *hapu*, or if either parent can establish a claim to title by filiation. It is significant that membership can apparently also be acquired by marriage to a member and taking up residence in the *hapu* territory. But presumably recruitment to *hapu* membership is predominantly from within. In this case membership is conferred by being born within the *hapu* of parents at least one of whom must (like Pericles's Athenians) have title to membership by filiation or the extended form of filiation documented by a pedigree.

There is, however, a second question with regard to the internal structure of the *hapu*. This is the question of how all the members of a *hapu* are linked with one another. They are obviously all linked with one another, no doubt in varying degrees of closeness, by local contiguity and their common interest in their territory and its defence. This is analogous to the common interest which all the residents of an English village have in the amenities and social services of their village. These links might be reinforced by hereditary land rights. But they are compatible with a community structure in which the members might well be grouped in discrete parental families some of which might be interrelated by affinal ties. The common bonds might then be a common interest and participation in a *hapu* cult or in a social service like a school, or in maintaining a communal economic asset like an irrigation system – and above all in the internal government of the community. But this does not seem to be the case. It seems that all the members of a *hapu* are related to one another by 'bilateral' and 'ambilateral' kinship and affinal ties, that is, ties through either or both parents and through either male or female antecedents. In my view, such an internal web of kinship is not sufficient to make a unit of political or territorial association into a bilateral descent group. It would have to be strictly endogamous to be that. But it might be that each individual member or sibling group can and must establish a pedigree, through either or both male and female parents and forebears, connecting him with a unique Founder-hero. Given the absence of *hapu* endogamy it means, of course, that there will be descendants of the Founder-hero in other *hapu*. Furthermore, since a Maori can be a member of a *hapu* through filiation on one side and yet can claim membership in another *hapu* through filiation on the other side, it may be conjectured that complementary filiation is a significant element of Maori social structure. It seems to play an important part in establishing land and property rights. This is one of the links, if not the major link, between the internal, kinship domain of the *hapu* and the external politico-jural domain of inter-*hapu*, that is tribal, relations.

An interesting problem is why 'ambilateral' kinship ties are so elaborately utilized in the internal structure of the *hapu*. The exceptional importance of hereditary rank undoubtedly puts a premium on capacity to establish a good pedigree. What, then, are the advantages of the flexibility of entitlement to rank which, as Firth shows, ambilateral kinship provides? The answer, I think, lies in the economic and political factors stressed in Leach's concluding remarks (*loc. cit.*, 1957). I would draw attention to only one such factor, the apparent absence of judicial institutions and sanctions, comparable, for example, to those of the Tswana, in the Maori social system. I infer this from the indications (Firth 1929, ch. XI, *passim*) that self-help was a normal means of redressing wrongs and trespass even within families. The widely diffused solidarity generated by the ramifying kinship ties within and even between *hapu* would, I suggest, be a valuable, possibly essential basis for social control in a political system of the Maori kind.

My argument can be summed up in this way. It is not enough to speak of descent or kinship or, in particular, of 'groups', in general. It is essential to specify the structural domain to which the analysis refers. Kinship or descent may confer a title to membership of a political or cult or economic 'group'. It does not make that 'group' into a kinship or descent 'group' in any absolute sense. As a matter of fact, I do not see how the concept of a 'descent group' is applicable in the conditions of 'ambilateral affiliation' described by Professor Firth, for the 'group' is never closed by a descent criterion. And I must confess that I am puzzled by the statement that there are systems 'in which the major emphasis is upon descent in the male line, but allowance is made, in circumstances so frequent ... as to be reckoned as *normal* [my italics] for entitlement through a female' (Firth, *loc. cit.*). It needs much amplification, I think. For example, it might be interpreted to mean that membership (of a rank-holding or property-holding, or otherwise corporate 'group') is strictly tied to 'the male line' – and therefore vested in males exclusively – but daughters (*sc.* sisters) have some kind of secondary or dormant rights of membership which they never forfeit and are able to transmit by filiation to their offspring. Then in certain circumstances (if the male line fails and there are structural requirements for the 'group' to be maintained) the dormant rights can be actualized in virtue of matrifiliation, and a member recruited into the 'group' as if he were a true descendant in the male line. Or it may mean that rank, for example, is transmitted equally through men and women by filiation to both sons and daughters, so that every person has title to both his father's and his mother's rank. If there are degrees of rank then (taking the inside view) a person might be allowed to choose in terms of his own advantage whether to align himself by patrifiliation or by matrifiliation with a rank 'group'. If there is a 'major emphasis' on the male line this must be from the outside standpoint of the whole social system. It must apply to a very special office or rank or quality deriving its validity from its significance in the politico-jural domain, not from within the 'group'. For if it were only an internally superior rank anyone who joined the 'group' by matrifiliation would be in an inferior position and would only join it if the advantages he derived were outstandingly greater than those he would have as a member of his father's 'male line' group. If it is purely a choice of rank, then some 'male lines' must be inferior to others and it must be more valuable to be a 'second-class' member of a higher-ranking male-line group than a 'first-class' member of a lower-ranking group.

These comments on the *hapu* apply equally, in my opinion, to the kind of land-owning groups described by Goodenough in the paper (Goodenough 1955) cited by Firth. The comparison that comes to mind is much more that of a joint stock company in which the individual's right of 'membership' rests on the acquisition by purchase or otherwise of stock than that of a lineage or other type of descent group. The difference is that in a politico-jural system without differentiated judicial and contract-enforcing institutions, and in an economy which lacks money and exchange institutions, title to 'stock' derives from filiation. In both cases it is the 'stock' – that is the estate – and the functions of the organization in the external politico-jural domain that make it a corporate unit, not the mode of recruiting members.

The foregoing discussion leaves a number of important theoretical problems open. Thus, I am not happy with a terminology, however qualified (as by Firth) that describes associations of the *hapu* type as 'descent groups'. But there can be no doubt that they are corporate political associations whose members are linked by common kinship ties. We might perhaps revive a traditional term and call them 'kindred groups'. This would emphasize the fact that they are not recruited on the principle of perpetual succession but on the more general principle of filiative kin right.

To sum up, kinship and descent rules are not only criteria for uniting people in groups and linking them in person-to-person relationships, but also and equally institutions by means of which status in the politico-jural domain is attributed and legitimized. The structure of that domain is therefore as significant for the way in which kinship and descent are utilized as are the irreducible facts of parenthood, siblingship and marriage.

Appendix to Chapters 3 and 4

In the interval since the publication of these papers, the topics dealt with in them have been voluminously discussed, as I have remarked in the Preface. The main issues that have thus arisen are examined at length in my forthcoming book. Here, I confine myself, therefore, to a brief consideration of the more important recent additions to the debate. During the past dozen years much of the ethnographic research that was originally made available to me in unpublished form has been published, and a wealth of new data has been added. This has played a big part in clarifying the conceptual problems I was concerned with in these papers.

The central theme of the debate has been the status and applicability of the concept of descent, and of the correlative concept of filiation, in the analysis of different types of kinship-ordered social structure. Scheffler's recent survey (1966) admirably delineates the differences of view that have been maintained and reaches conclusions with which I largely agree. The queries raised by Leach (1957 and 1961) which led me to re-formulate the main points of my 1953 analysis in the paper here reprinted as Chapter 3, remain relevant. Thus one question that has been in the forefront is that of the sense in which the concept of descent can be appropriately applied in the analysis of so-called 'non-unilineal' or 'bilateral' or, more generally 'cognatic' kin group. Some authoritative writers, including Goodenough (1955), Davenport (1959; 1963), Firth (1957; 1963) and Murdock (1960) claim that groups like the Maori *hapu*, the Mangaian *kopu*, the Gilbertese *oo*, etc. are properly describable as 'descent groups' based on the same fundamental principles as the strictly unilineal descent groups of African lineage theory. The criteria emphasized are their recruitment by credentials of kinship traced to ancestors, their continuity, and their corporate constitution.

Those of us who would prefer to restrict the term 'descent group' to the unilineal type I discussed in my 1953 paper (*cf.* Goody 1961, Leach 1962 and Scheffler, *loc. cit*) consider this to be confusing. The point is that these 'cognatic groups' are not marked off from and counterposed to like groups in the total society by a uniform descent boundary that strictly and predictably defines membership, nor even by a boundary set in purely kinship terms. It is true, that, considered from within, they are recruited, and comprise persons connected, by kinship ties or marriage which may be documented in individual pedigrees. But the groups that are externally demarcated from one another – that is, not the ramifying personal kindreds also found in these systems – are in fact segregated by virtue of political, or territorial or economic (e.g. land owning) or ritual boundaries, not by reference to exclusive kinship or descent criteria.

The confusion is due primarily to using the term 'descent' to cover all modes of recognizing ancestry for social purposes, so long as it can be documented by a pedigree and then transposing the concept to designate any kind of 'group' in which membership is in any way contingent upon demonstrated or putative ancestry (*cf.* Scheffler, *loc. cit.* and see also Fried (1957)). Restriction of the term 'descent group' (but not necessarily the concept of descent) in the sense originally suggested by Rivers (1924) at least has the merit of conducing to conceptual clarity and of corresponding to distinctions recognized by the actors themselves.

The ambiguity is avoided, I suggest, if we define 'descent' by contraposition with 'filiation' rather than in terms of its dictionary meaning. This is, I believe, closer to the actor's conceptualization and system of norms than the blanket usage of common parlance. It corresponds, also, to the division, widely represented in modes of residence and in custom and law, between the domain of familial relations and the domain of politico-jural relations. And it enables us to distinguish between credentials of right and duty based on different ways of reckoning connections with ancestors lineally or collaterally, by steps of filiation. The analytical value of these distinctions is brought out in Barnes's discussion of the role of 'cumulative filiation' as opposed to unilineal descent, in New Guinea Highland societies (Barnes 1967). The field researches referred to by Barnes, and more recent additions to this important body of new data by Meggitt (1965), Brown (1962), Langness (1964) and others, have shown how 'dogmas of descent' may be used to identify segments of tribal structure, recruitment to which is, in fact, based on a variety of filiative and affinal kinship credentials.

The concept of filiation is now in common circulation but its implications obviously need more clarification. Firth (1954) and

Leach (1962) have pointed out that complementary filiation serves the purpose of providing for individual choice in some areas of social alignment in systems with unilineal, corporate descent groups. Freeman's definitive analysis of the 'kindred' in cognatic systems (1961) shows brilliantly how such bilateral constellations of actor-centred kin are constituted by the selection and combination of filiative connections. One study of special interest is Robinson's re-examination of the role of the father in Trobriand marriage (1962). It is clear that complementary patrifiliation is recognized as entitlement to specific rights and privileges relating to a daughter's marriage and this is demonstrated in the gifts and prestations exchanged.

Filiation is the main topic of a collection of papers introduced by Schneider (1967). Seeking to elaborate further the conceptual distinction between descent and filiation, Schneider reconsiders the symbolization of the variables thus subsumed, to which brief allusion was made in my paper of 1953. He is at pains to emphasize the 'cultural' significance of the proposition that a 'person as kinsman' is made up of 'a number of elements' (*loc. cit.*, p. 68), in a parallel sense, I take it, to 'an assemblage of statuses', as I originally called it (*cf.* above p. 92). One of the papers appended to Schneider's introduction (Sider 1967, pp. 90–109), is of interest for the confirmation it provides of Robinson's conclusions about Trobriand patrifiliation.

The contraposition of descent and complementary filiation figures in a different way in Goody's proposal for classifying double unilineal descent systems (1961). Insisting on restricting the concept of descent to the unilineal category, he separates unilineal systems with complementary descent groups which are non-corporate, from those in which both categories of descent groups are corporate, thus bringing conceptual order into this hitherto confused subject.

The question of what constitutes a 'corporate group' has received much attention. The view that the critical feature is the possession by the group and transmission within it of specific kinds of property, notably land and similar productive assets, or, in more general terms of an 'estate' has authoritative advocates. (*Cf.* Fried 1957; Firth 1959, 1963.) It has been advanced with great learning and ethnographical insight by Goody (1962). Freedman (1966) writing of South-East China, and Gough (1961) in her description of Nayar lineage organization, have used this criterion to distinguish between the 'corporate lineage' of relatively recent common ancestry and the higher levels of non-corporate lineage organization ordered to rules of exogamy and of ritual observance. M. G. Smith, however (1956), supports the proposition (*cf.* p. 78 above) that the essential feature is the conceptualization' of the corporate groups as 'one person'. In this I concur, with the proviso that 'one person' must be understood in the politico-jural sense equivalent to the notion of the 'juristic person' of Jurisprudential theory. I consider that the balance of the ethnographic evidence is in favour of this interpretation.[1] Whether or not it is explicit in the laws of property or the definition of status, or implicit in the mode of genealogical validation or in the symbols and customs expressive of common identity, this is what lies behind the perpetuity of the 'corporation aggregate' and of its correlate the 'corporation sole', as Maine long ago (1861, ch. vi) made clear and as Smith cogently argues (*loc. cit.*, pp. 64–67).

Those who claim 'non-unilineal descent groups' to be functionally equivalent to unilineal descent groups cite, in particular, their purported corporate character (*cf.* Firth 1963), implying that corporateness and descent are linked together in both cases. However, others acquainted at first hand with societies in which cognatic kinship is the basis of social organization, think differently. Thus Leach (1961, p. 7) declares locality and not descent to be the 'basis of corporate grouping' in Pul Eliya. The debate continues and will doubtless go on as long as ambiguous conceptions of the nature of descent and of the corporate group persist.

Finally, segmentation. Southall's extension of the concept to describe the hierarchical localization and distribution of political authority among the Alur (1954) gains support from M. G. Smith's re-examination of the connection between lineage structure and political organization. Smith's dictum (*loc. cit.*, p. 54) that 'all political organization involves segmentation' and his inference that no distinction can therefore be drawn between 'lineage societies' and those which are not 'organized on segmentary principles' to my mind, however, go too far. A

---

[1] A notable ethnographic example is Goodenough's perspicacious study of Truk economy and social structure, 1951.

Tswana age-regiment is not a political 'segment' of the same structural form as a Nuer lineage.

The problem of segmentation, and the associated processes of fission and fusion, accretion and splitting, form the central theme of the exemplary study by Meggitt to which I have already referred (1955 and see Barnes 1967). Marshalling an impressive array of quantitative evidence, Meggitt analyses the processes of recruitment to the formally agnatic clans by patrifiliation, matrifiliation and the incorporation of collateral kin. He shows how demographic variations over time and pressure on land contribute to the changing patterns of segmentation in the system and relates this to their corporate structure and functions in economic, matrimonial, funerary and war-making customs and practices. The conclusion he reaches by a comparative survey of New Guinea Highland social systems is that 'where the members of a homogeneous society of horticulturalists distinguish in any consistent fashion between agnates and other relatives, the degree to which social groups are structured in terms of agnatic descent and patrilocality varies with pressure on available land resources' (*op. cit.*, p. 279). This conclusion is, however, strongly contested by other authorities. Barnes, for example, points out that there is no consensus of judgement on criteria for determining 'degrees' of agnation or matriliny (*loc. cit.*, 1967). He concedes that the Mae Enga 'segmentary system is undeniably a political and jural phenomenon' (*loc. cit.*, p. 4) but questions the validity of describing it as an agnatic system. What, in short, Meggitt's and other recent researches reveal is complexities in the empirical data we are accustomed to categorize by such concepts as agnation, matriliny, and so forth, which demand further discrimination of the variables thus subsumed.

This lesson has also been emphasized in Schneider's learned and trenchant essay (1965) on the supposed divergence between 'descent theory' and 'alliance theory'. It is not, he makes us realise, a question of mutually irreconcilable 'models' of the same empirical phenomena but rather one of different theoretical interests directing attention to different, in this case complementary, dimensions and constituents of systems of social relations ensuing from the combination of kinship, descent and marriage institutions.

5

Analysis and Description in Social Anthropology[1]

Presidential addresses to this section of the British Association have been some of the most notable landmarks in the development of the anthropological sciences. This is not surprising when we recollect the names of those who gave them. Haddon (1905) and Rivers (1911), Seligman (1915) and Radcliffe-Brown (1931) could not fail to make important additions to knowledge and advances in theory on the occasion of a presidential address to Section H of the British Association. But in science and scholarship, unlike letters and the fine arts, advances made by leading personalities are much more the decisive steps in a collective effort than the fruits of pure inspiration. Did not Isaac Newton say that he stood on the shoulders of giants? When we look back to two such seminal events in the history of social anthropology as Rivers's presidential address of 1911 and Radcliffe-Brown's of 1931, we see how true this is. Entangled in the controversies of the day, Rivers thought he was defending the so-called historical point of view which was so soon to fizzle out in the diffusionist extravagances of Elliott-Smith and Perry. Yet he was already being swept into the current of ideas which brought about the revolutionary transformation of social anthropology summed up in Radcliffe-Brown's address of 1931. He was struck by what seemed to him to be the remarkable observation that the indigenous social structure of a people often maintains itself in the face of greatly changed political conditions, and in spite of the disappearance of their material culture and their religious customs and beliefs. Ten years later the current of thought he was indirectly responding to took massive shape in the first monographs of Malinowski and Radcliffe-Brown (1922); and if one wanted to

[1] Presidential Address, British Association for the Advancement of Science, Section H, 4 September 1953 (reprinted from *The Advancement of Science*, No. 38).

sum up briefly one of the chief results of the field research inspired by Malinowski and Radcliffe-Brown, one could say that
it lies in the understanding we have now reached of the full
implications of Rivers's remark. In undertaking the office you
have honoured me with I put my trust in the precedent I have
just illustrated. Never before has social anthropology in this
country been so vigorous and productive as in the post-war
decade just drawing to a close. My task will be well fulfilled if I
can present some aspects of this collective enterprise.

Radcliffe-Brown's presidential address of 1931 was memorable
for the formulation he gave in it of the principles of theory and
method which have inspired the best field work in social anthropology during the past thirty years. In his own words, this method
and theory 'looks at any culture as an integrated system and
studies the functions of social institutions, customs and beliefs
of all kinds as parts of such a system'. The triumphs of ethnographic field work during the past thirty years have been the
result of approaching the tasks of description and analysis of
field data in terms of this principle.

We can see what a difference this approach has made to social
anthropology since Rivers's day if we compare his justly famous
study of the Todas (Rivers 1906) with the first, and still un-
rivalled masterpiece of functionalist ethnography, Malinowski's
Argonauts of the Western Pacific. The comparison is instructive
because if any ethnographic study can be regarded as a precursor
of Malinowski's it is surely that of Rivers. Did he not state, at the
very beginning of his book: 'The whole of Toda ceremonial and
social life forms such an intricate web of closely related practices
that I rarely set out to investigate some one aspect of the life of the
people without obtaining information bearing on many other
wholly different aspects' (p. 10)? But instead of seeing this as a
discovery calling for a new way of regarding custom and social
organization, Rivers thought of it as an impediment in his task
of separating out the different aspects of social life. He thought
that scientific method required him to keep apart the description
of the facts of observation and the theories by which he explained
the facts. He dealt with the 'facts' by enumeration and description;
his explanations or theories were tacked on to them and were
couched in the idiom of the time. So, to take an example at
random, he interprets the fact that a woman after child-birth

undergoes ritual associated with the dairy as possibly relics of a time when women had more to do with the dairy than at present (p. 330). For Malinowski the inter-connectedness of the customs and institutions of the Trobrianders was the discovery which determined his handling of the observed facts; but in addition, following a more sophisticated view of scientific method, he refused to separate fact from theory. Every way in which facts are grouped in description involves theories, implicit or explicit, about the connections between them that are significant; and significance is a function of the kind of questions to which the observer seeks an answer. The reality of social life in any society is a welter of human activities, material objects and natural events. Science, which Norman Campbell, I believe, somewhere calls the greatest of the arts, works by selecting from the welter and seeking to establish relationships between the data selected. Criteria of relevance, rules of procedure and forms of communication are pre-supposed in this. And behind the selection, arrangement and presentation lie the kind of questions asked by the investigator (cf. Kaufman 1944, ch. v; and Nadel 1951, passim).

The new ethnography introduced by Malinowski and Radcliffe-Brown is quite explicit on this point. Ethnographic facts are meaningless, except in the colloquial sense of compatibility or incompatibility with the observer's commonsense categories, unless they are examined in the light of theory. The ethnographic monograph of today, as Professor Evans-Pritchard points out (1951, ch. v), is generally a study of a problem – that is, an investigation of an hypothesis – as well as a record of field data. Of course the ethnographers of the past, for all their pretended neutrality in matters of theory, in fact held quite definite theories as I have already suggested. Every general term they used, even such innocent looking ones as 'family' and 'clan', was loaded with theoretical assumptions. But the deliberate effort to make a field study both an investigation of hypotheses and a means of arriving at new hypotheses that is our practice today, is a recent development.

This is one of the chief features that distinguishes functionalist ethnography from its precursors. It is epitomized in the reproach often cast at Malinowski, that he generalized about humanity at large from his experiences among the Trobrianders; but criticism has not stopped his successors from continuing in this habit. And one reason for this is that the results have handsomely vindicated the procedure of testing, amending and adding to the generalizations which make up the body of social anthropological theory by intensive study of one society at a time. For it is a method more in line with some experimental procedures in the natural sciences than any of the versions of the comparative method, which is commonly supposed to be the best way of reaching valid generalizations in social anthropology. It resembles the kind of experiments made with a single species of plants or animals classically typified by Mendel's work on sweet peas.

Functionalist ethnography has been so fruitful in matters of theory because the principles which guide it in looking for significance in ethnographic observations require an analytical, as opposed to a descriptive mode of investigation. It is worth adding that comparative studies which have been of theoretical value, not just accumulations of parallels and illustrations, likewise follow analytical procedures. This was brought out in Professor Forde's presidential address to this section (1947), and the example he used, 'the character and role of kin groups in social organization', is particularly apt. It is obvious in Professor Radcliffe-Brown's recent defence of the comparative method (1952). Comparison is a secondary step in establishing a generalization. Seeing that the number of human societies, past and present, is finite and small, we can, by comparison, establish the range within which a proposition concerning a general tendency is valid; the hypothesis itself is commonly derived by the study, in the first instance, of customs, institutions, or social relations in a single social system.

To explain what I mean by analytical methods let me first note that, in order to distinguish units of custom or of social organization, the ethnographer has to isolate standard and recurrent features of social situations (cf. Nadel 1951, ch. v). When I say that among the Tallensi a first-born is prohibited from eating fowl I am making a statement based on the analysis of a representative sample of information about first-borns and of observations of their behaviour in appropriate situations. But this level of analysis is fairly elementary, though too often taken for granted without more ado, as I have explained elsewhere (1949). For practical purposes it can be regarded as description; and though every analytical step in ethnography opens up questions of theoretical significance, the higher the level of analysis the more important is the theoretical problem thrown up, since the level of analysis is related to the diversity of the descriptive data referred to.

The essence of description is that observations are grouped together in accordance with their actual relationships and contexts of time and place. Ethnographic literature is so largely descriptive that it might seem superfluous to give examples, but as I am going to refer to marriage customs and institutions later on, it may be useful to remind ourselves how these are dealt with by the method of description. Marriage is described as a sequence of customary activities following one another in a set pattern. The description generally begins with some account of heterosexual interests in children before they reach marriageable age. There probably follows a brief reference to initiation or puberty ceremonies where these occur; then a description of the normal customs of courtship and betrothal, with an aside about the proprieties regulating social relationships between the different parties during the betrothal period. There follows an account of the wedding rites and celebrations, and finally a discussion of the legal formalities such as the payment of bride price. To round off the account there may be some consideration of divorce and other ways of terminating marriage.

This formula is filled out in various ways by the addition of graphic detail and commentary but it has remained essentially unchanged for fifty years. It is sometimes given an air of theoretical respectability by calling marriage a *rite de passage* or by lumping it in with kinship institutions. While this brings marriage into relation with the wider field of social organization of which it is a component institution, it still follows the sequence and concomitance of things as they actually happen. Marriage is a stage in the journey through life; or it is the prelude to the setting up of the family. The approach is that of the natural historian.

Analysis takes a different course, and its nature is most easily seen when it occurs, as commonly happens in ethnography, in a descriptive context. For we can see then that its aim is to find answers to questions of a different order from those that receive descriptive answers.

Analytical questions refer to what Whitehead called 'ideal isolates' by contrast with the heterogeneous welter of empirical data; and ideal isolates are theoretical constructions. The analytical method is to break up the empirical sequence and concomitance of custom and social relations and group the isolates so obtained in categories of general import. As it is an indispensable step in the movement from field observation to theory and back to the test of further field observation, it has often been implicitly followed by anthropologists ever since field research began.

Description cannot yield generalizations; we can arrive at generalizations only by way of analysis. But for this, the ideal isolates we use must have meaning in terms of the descriptive reality of social life, and the isolates we use in the analysis of one class of anthropological facts must be consistent with those we use in the analysis of any other class. In other words, our isolates must form a theoretical system. It is through positing isolates that have no counterpart in social reality, or are inconsistent with the general body of valid theory, that many plausible anthropological hypotheses have come to grief. Advances in social anthropology depend on devising the right kind of isolates. Rivers was an acute observer. But he confounded the isolates he worked with in his most ambitious theoretical study, *The History of Melanesian Society* (1914). He correctly perceived that 'distinctions in (kinship) nomenclature are ... associated with distinctions in conduct' (vol. 1, p. 45). But instead of following up rigorously this synchronic correlation, as Radcliffe-Brown so profitably did, he introduced a contradictory and empirically useless isolate in the form of the hypothesis that kinship nomenclatures contain 'anomalies' which are relics of extinct marriage and sex customs.

The most promising tendency in social anthropology today lies in the development of analytical methods and isolates within the framework of the functionalist hypothesis that the customs and institutions of any people make up a system of interdependent parts and elements, which work together to maintain the system in a steady state and have value for the realization of legitimate social and personal goals.

I want to emphasize that customs and institutions are not isolates but descriptive units. When we speak of 'the family' or of 'sacrifice' we are referring to descriptive not analytical units. But when, to quote two rather different classical examples, Malinowski wrote of 'reciprocity as the basis of social structure' (1932, ch. IX) and Radcliffe-Brown discussed the functions of kinship institutions in distinguishing, ensuring and transmitting different kinds of rights over things and in persons (1935) they were developing isolate concepts. This is clear from the fact that Malinowski, for example, tried to establish the validity of his generalization by examining a range of descriptively diverse units of custom, and claimed that it held over 'the whole culture and the entire tribal constitution' (1932, p. 49). Isolates are abstractions (cf. Evans-Pritchard 1951) and it is characteristic of abstractions that they may vary greatly in generality. The isolate 'lineage' discussed in Professor Forde's presidential address is of less generality than the isolate conceptualized in the phrase 'rights in personam' as used by Radcliffe-Brown, which may include rights of a chief over his subjects or of a priest over adherents to a cult.

The argument can be put simply in this way. In analytical terms we think of a custom, an institution, a belief, a social relationship, any unit of actual social life, as if it were made up of a number of isolates, and we think of an isolate as if it were an element that enters into the formation of different kinds of customs, institutions and social relations within one society. And our isolates must form a theoretical system because we have to define them in relation to one another. Lineage descent is defined by distinguishing it from such other isolates in the same system as kinship filiation; rights in personam are defined by contrasting them with other kinds of rights and with their negative, the absence of rights. Furthermore, our isolates must pass the fundamental test of a good theory, which is that it makes possible generalizations that bring together and explain empirical observations not previously seen to be related to one another. Such very general concepts as that of the integration of a social system or that of the consistency of a people's culture are meaningful as isolates at a high level of abstraction.

However, there is no argument so strong as a good example, and this is what I want to turn to. Recent studies of marriage in African societies are particularly illuminating. Since a survey of the data has just appeared (Philips, et. al., 1952) I will not attempt a general review but use one or two specific examples to illustrate my argument. Theoretical discussion of the institution of marriage in Africa has centred mainly on the widespread custom by which a payment is made by or on behalf of the husband to the wife's kin. Thirty years ago Junod in his well-known ethnography of the Thonga (p. 121, vol. 1, 1927 edition) explained this custom as follows:

> To understand them, we must observe that marriage in primitive or semi-civilized tribes is not an individual affair as it has become with us. It is an affair of the community. It is a kind of contract between two groups, the husband's family and the wife's family. What is the respective position of the two groups or families? One of the groups loses a member, the other gains one. To save itself from undue diminution, the first group claims compensation, and the second grants it under the form of the lobolo. This remittance of money, oxen or hoes will allow the first group to acquire a new member in place of the one lost, and so the balance will be kept. This conception of the lobolo as a compensation, a means of restoring the equilibrium, between the two groups is certainly the right one.

This explanation, with modifications, reappears in many subsequent analyses of marriage, especially in African societies which have corporate patrilineal lineages. Professor Radcliffe-Brown, for instance, in his recent exposition of his theory of kinship and marriage (*Introduction* to Radcliffe-Brown and Forde 1950) agrees that from one aspect 'the marriage payment can be regarded as an indemnity or compensation given by the bridegroom to the bride's kin for the loss of their daughter' (p. 50), and he then refers to those societies – of which there are many in Africa – where the daughter's bride-price is used to marry a wife for her brother or some other male of her natal family. But, as he emphasises, this accounts for only one aspect of African marriage payments. Another and often more important aspect is that marriage signifies a transfer to the husband and his kin of certain rights in relation to the wife and the children she bears. Thus the full meaning of formalities carried out at the time a couple begin to cohabit and set up house is apparent only in terms of the structure of the family that results some years later.

The idea that marriage in primitive society is basically a form of exchange derived in the last resort from the general principle of reciprocity, is also developed with great learning and brilliance by Professor Lévi-Strauss (1949); and though his primary interest is in cross-cousin marriage he extends his hypothesis to include marriage of the type common in Africa. It is not possible to do justice to the subtlety and many-sidedness of Professor Lévi-Strauss's analysis in a few sentences, but for purposes of my present argument it is enough to pick out his contention (p. 582) that bride-price payment is a means of bringing about deferred and roundabout exchanges of women for women in a society made up of many exogamous groups.

It is unnecessary to list all the writers who have based ethnographical or comparative study of African marriage essentially on the indemnity (or exchange) hypothesis for the methodological point to be obvious. Descriptively, it is undoubtedly the case, in patrilineal societies at any rate, that one 'group of kin' (not further specified for the moment) 'loses a woman' and through the medium of the bride-price 'recovers a woman'. Analytically considered it is not so straightforward, as Radcliffe-Brown shows by the qualifications he adds to the exchange hypothesis. The transfer of rights in a woman, let alone in her as yet unborn offspring, is a different thing from an exchange of woman for woman. The concept of rights refers to an isolate, to a special kind of social relation between two parties which is enforceable by sanctions. As soon as we put it this way we see how different is the level of analysis from that involved in reducing marriage to a special form of exchange. The patient observer can almost literally see bride-price being exchanged for a woman. To arrive at 'rights' he has to examine the whole range of social relations in which the married couple have joint and separate roles and also consider marriage as a developing process through a stretch of time.

This is well brought out in a paper by Dr Laura Bohannan (1949) which sums up the analytical method of handling this problem. In this paper Dr Bohannan examines Dahomean marriage with a view to finding out if there are any common principles behind the thirteen different 'types' described by the ethnographic authorities. She concludes that they are all different ways of arranging the distribution of rights in a woman, the two kinds of rights at issue being those held in a woman as wife (*in uxorem*) and those held in her as to the children she may bear (*in genetricem*). These rights can be conjoined, or can be vested separately. In one type of marriage, for instance, which would descriptively be classified as a 'matrilocal' variant of the more normal 'patrilocal' arrangement, what analysis shows is that only rights *in uxorem* are transferred to the husband in return for the bride-price, the wife's lineage retaining rights *in genetricem* –

that is, jural control of the offspring of the woman thus married. In the case of the royal lineage rights over the children of daughters are never transferred but the women are given in marriage, with transferred rights *in uxorem*, to the King's political favourites. This explains how in Dahomey, as in many other parts of Africa, a woman of wealth can 'marry' a woman. What happens is that the wealthy woman acquires rights in the reproductive powers of a young woman and hands over rights in her as wife to the man who is brought in to father the children. What is specially interesting is that the bride-price gives a man rights only over the marital and domestic services of his wife; a ritual act in which the lineage ancestors are informed of the marriage is necessary to confer rights over the offspring. It is significant also that the distribution of these rights *in personam* is closely related to the various ways in which rights of inheritance and succession are distributed and vested.

When we look again at the descriptive data in the light of these analytical concepts many things that otherwise seem not to be of great moment take on significance. Thus even in patrilineal societies in Africa a woman is never completely severed from her natal family and descent group by marriage. As I have described in detail for the Tallensi (1949) and others have shown for other African peoples, a woman never forfeits her status as daughter in her patrilineal lineage and family; but rules of residence, reinforced by very strong sanctions of the ancestor cult ensure that this status is never confounded with that of wife. To put it in terms of the reciprocals of Dr Bohannan's isolates, a woman's rights as daughter (*qua filia*) are not extinguished by the rights she acquires as wife and mother (*qua uxor et mater*). This is one of the reasons why the mother's brother plays so important a part in the kinship systems of peoples with patrilineal lineages. In short marriage is not an arrangement for exchanging women but a process of transferring – or perhaps exchanging – rights, the parties and commodities involved being each a focus for a cluster of rights which do not necessarily remain combined.

This initial step in analysis opens up a number of interesting possibilities. What we isolate as a right in jural terms we see as vested in persons, as embodied in structural relations, and as symbolically expressed in ritual beliefs. Or we can link this chain of isolates in the opposite way and end by asking how rights are

sanctioned. We are then led to the discovery, excellently illustrated in Professor Mayer's analysis of Gusii bride-price (1950),
that the bride-price has an element of a liability to those who
receive it, since it is at their service only in proportion to the
degree to which the woman for whom it is given adequately
fulfils the obligations of wife and child-bearer.

But we must turn to societies in which rights in property,
persons, political office and cult are vested in matrilineal descent
groups to see how fruitful analytical methods can be by comparison with a purely descriptive approach. A number of studies
of marriage in African societies in which matrilineal descent is
the critical factor in jural and ritual relations have been published.
I would draw attention, in particular, to Professor Forde's monograph on Yakö marriage (1941) and to Dr Richards's important
account of Bemba marriage (1940) as well as to her more recent
analysis of what she has aptly called 'the matrilineal puzzle' in
Central Bantu societies (1950). In the latter paper Dr Richards
discusses some of the researches of Belgian scholars, who have
devoted much effort to the study of marriage among the matrilineal peoples of the Congo. For my present argument, however,
it will be more satisfactory to use my own field data from Ashanti,
rather than to rely on these or other published studies. I have
referred to Professor Forde's and Dr Richards's papers because
my analysis has much in common with theirs.

The late Dr R. S. Rattray compiled the most important
descriptive facts about the customs and institutions of Ashanti
marriage in his great series of monographs more than twenty
years ago (1923-9). In the rural areas of Ashanti these customs
and institutions are unchanged in essentials. But if we want to
understand Ashanti marriage analytically it is not enough to
consider only the sequence of customary acts that normally lead
up to a marriage. The relevant isolates cannot be properly evalu-
ated without taking into account the whole range of institutions
involved in lineage structure, domestic organization and kinship.
I can only give some indications today.

Ashanti marriage is singularly lacking in the vivid and symbolic rites and ceremonies so common among Bantu peoples.
It is a purely jural transaction. But certain features of this transaction are the result of the Ashanti concept of the personality as
expressed in ritual notions. Every Ashanti belongs by right of

birth to his mother's matrilineal lineage. This automatically confers rights of citizenship and domicile in a chiefdom, as well as rights of inheritance and succession in matters of property and office vested in the lineage. The Ashanti lineage is a localized group of great solidarity, the men and the women ranking equally in economic and domestic affairs but the women being excluded from direct participation in the religious and political activities that are tied to the lineage. This solidarity is symbolized in the very stringent prohibition of sex relations between members of a lineage, breach of which is regarded as a sin and a crime and was formerly punished by death. The corporate unity of the lineage is the fundamental principle of Ashanti law, morals and religion (Fortes 1950).

Lineage membership gives the individual unconditional rights to use productive resources held by the lineage, to seek any office vested in it, to claim legal guardianship from it, and to have the ritual protection of the ancestors and gods of the lineage. These imply the right to assistance from the lineage in distress, to share in its prosperity, and most important of all, to be given a proper funeral by it. Correspondingly, the lineage as a corporate body has the right to the loyalty of its members in political and legal matters, to economic contributions from them in a common cause such as the funeral of a member, and to unconditional support in maintaining the resources, rank and reputation of the lineage. The individual cannot abjure membership of the lineage for without it he is jurally a nonentity. But the lineage can, with the unanimous consent of all members, expel a member who has consistently flouted lineage obligations, in particular a member who has infringed its moral solidarity by committing incest or seducing the wife of another lineage member.

The strength of a lineage lies in the number of its members. Clearly, therefore, its most vital asset is the reproductive powers of its female members. Control over these cannot be alienated, and this is a decisive factor in marriage. But there is another side to this matter. Paternity carries a very high value. A free person is born legitimate, that is, with the rights arising from lineage membership, no matter how or by whom he was begotten. He derives no legal rights from his father. And yet, it is both a disgrace and a disability not to have a father who has properly acknowledged paternity, even if he has not brought one up, as a father should. In descriptive terms this goes back to the Ashanti conception of the personality, to which I will return in a moment. Analytically it is connected with the great dichotomy in Ashanti life between all the social ties that spring from matrilineal kinship and all those that spring from the association of a man and a woman in marriage. The former only become fully effective in adulthood, of which the most important sign is socially recognized fitness to marry; the latter are of great weight for childhood, when education and character formation are taking place. The former are mediated by the sibling bond between mother and mother's brother and are focussed in the jural authority of the mother's brother; the latter are created through the conjugal bond of mother and father.

The Ashanti conception of the personality is in part a reflection of this dichotomy. Very summarily, every person has in his make-up a spiritual or temperamental element which comes to him from his father at conception and is symbolized in an hereditary ritual bond shared by close patrilineal kin. He also has an individual character that is partly a matter of fate and partly of up-bringing; and finally he has the jural personality derived from his lineage status. A child does not thrive if his father's soul does not watch over him with care and affection nor can he become a person of worth and virtue if he is not taught the proper skills, manners, morals and beliefs by his parents during childhood. Thus a child cannot grow up in the right way if he has no father to care for him (see references in Fortes 1950).

Conversely, a man's life is unfulfilled if he has no children. As a member of his lineage his concern is to have sisters' sons; as a man his chief aim in life is to have children. That is why impotence or sterility is as great a humiliation for a man as barrenness for a woman. The jural expression of this attitude is the father's right to give his child the name by which it will be distinguished from other members of its lineage. The moral expression of this is the child's duty to contribute to the support of the father in his old age, and, especially, to share with his other children the expense of providing his coffin when he dies. It is as disgraceful for an Ashanti to die without children to buy his coffin as it was for a Victorian Englishman to have a pauper's funeral. Last, but not least, there is the strength of paternal sentiment which is so often shown in the gifts *inter vivos* made by a father to his children (see Rattray 1929). One further item of descriptive information is pertinent. Ashanti do not like marrying outside their native town or village. They give many reasons for this but its most important effect is that husband and wife can, if they live in the same village, continue to reside with their respective maternal kin without detriment to their marital relationship. It is quite common for them not to have a household of their own (see Fortes 1949). This graphically documents the hold of matrilineal kinship on the individual and the fact that the conjugal bond straddles two lineages.

We can now consider the formalities necessary for a proper marriage in Ashanti and see what they mean (Fortes and Kyei). They consist of gifts, formerly often in kind – palm wine, antelope meat, cloths, etc. – now generally in money. The gifts pass, in the usual African way, from or on behalf of the husband to the bride's kin and the bride herself. There are customary standards of appropriateness for these gifts but the amounts vary according to the circumstances of each case. The tendency is for the amounts to be larger the higher the rank or social standing of the bride's maternal kin or father, but they are never very large. The kinsfolk involved are the father and mother, the lineage head and lineage elders, sometimes the brothers, of the bride and bridegroom. Friends of each family, who may or may not be kin, also play a significant part.

Two conditions are essential before any of the gifts that are jurally critical can be offered. Firstly, the bride must have undergone the public ceremony of declaring her nubility which all girls formerly went through on reaching the menarche and most girls who have not been to school still go through. This marks her attainment of adulthood. It is a sin and a crime, as heinous as incest, for a girl to conceive before this ceremony. Both she and her lover are driven out of the village. The parallel with incest is instructive. The point is that a woman's reproductive powers which come with sexual maturity, are not her own to dispose of at will. The community recognizes the existence of these powers and her maternal kin proclaim their control of them.

The second condition is that the girl's parents and lineage elders, especially the latter, must consent to the marriage, and, in the ideal case, the boy's too. On the jural side this is necessary for the gifts to be given and accepted by those who have the rights *in personam* which empower them to do so in binding form. On the personal side this enables the girl's maternal kin to assure themselves that her offspring, their future heirs and successors, will be worthily fathered, and the man's paternal kin to assure themselves that their names and spiritual affiliation will be passed on to children of a virtuous and reputable lineage. It is worth noting that, though he has no jural powers in the matter, the consent of the father is particularly desirable for both spouses, and records of several hundred extant marriages show that it counts almost as much as the consent of the mother's brother, who represents lineage authority in this context. This is a recognition of the spiritual affiliation a person has with his father and of the moral debt due to him for educating and bringing up his child.

The formality which makes a marriage legal is the passage of what is usually translated as the 'head rum *(tiri nsa)*'. This is a gift of drink, either palm wine or imported liquor, or cash in lieu. It is often described as a thanking gift *(aseda)*. Drink is given in exactly the same way by the recipient whenever rights in property or office or persons are ceded or conferred. It marks the transfer, before witnesses, as final.

In marriage, the drink is taken by representatives of the bridegroom's lineage head and elders to the head and elders of the bride's lineage. The latter sends half to the bride's father. The remaining half is ceremonially shared by the representatives of the two lineages and the bridegroom declares that he accepts responsibility for any debts his wife may incur and agrees that treasure trove she may find belongs to her lineage. The marriage is now jurally ratified. Judging by a sample of nearly 300 extant marriages contracted over a period of thirty years (Fortes and Kyei) this formality is carried out in 90 per cent. of marriages. In the remaining 10 per cent. of cases no formal gifts whatever have passed; but they include a number of couples who have, with the consent of their respective kin, lived as husband and wife for many years and have several children.

The difference between a proper marriage and cohabitation is that in the former case the husband is entitled to damages from any man who commits adultery with his wife, and will also receive an apology from her and her lineage; in the latter the husband has no redress if his wife is unfaithful. Again, in marriage with 'tiri nsa' the husband can insist on his wife's carrying out such domestic services as cooking and helping him on his farm; in the latter he cannot. So cohabitation can only be a stable arrangement if the spouses and their kin have great confidence in one another.

Where 'tiri nsa' has been paid the husband's paternity of the children of the union is taken for granted, since he is their presumed physical father, but he must all the same give the customary gifts and perform the naming and other ceremonies by means of which he accepts and claims paternity. A man establishes paternity in exactly the same way in a cohabitation union or, if he wishes, in the case of a child born of a casual liaison.

To complete the picture I must add that divorce is easy and common at all stages of marriage and parenthood, and case histories show that it is more often the wife than the husband who initiates the action. The main grievances are incompatibility, neglect of domestic duties by the wife, insufficient support by the husband, and gravest of all, sterility or impotence. A divorce must be formally ratified in the same way as a marriage, only in reverse. The wife's lineage head returns the 'head rum' to the husband's lineage head and elders with the same ceremony as took place when he accepted it. Whether or not there are children of the marriage is irrelevant.

Two other gifts from the bridegroom are of interest. One is the gift of a sheep or its money equivalent to the bride's father. This is for him to sacrifice to his soul, which has kept spiritual guard over the girl from childhood and is now asked to bless the marriage. The other is a small gift to the girl's mother to console her for the loss of her daughter's companionship. These are non-returnable gifts.

Finally, there may at any stage of the couple's life together be a transaction of a different kind. The husband may be asked to make a loan of any amount called 'head money' (*tiri sika*) to the wife's lineage. It is found in about one-third of extant marriages. It is actually a kind of mortgage, the security being the wife's fidelity. The wife has a stake in the welfare of her lineage. By consenting to the loan she pledges herself, both to her husband and her lineage, to perform faithfully her duties as a wife. Repayment can only be demanded if failure on the wife's part leads to divorce or when she dies; and if the wife has been dutiful and faithful for many years, or till death, the husband may make a free gift to her or her lineage of the loan. Whereas the 'head rum' provides no sanctions on a wife's conduct, 'head money' exploits the wife's lineage bonds to bolster up her marital loyalty. This is neat evidence of the normal condition of mutual exclusiveness, if not opposition, of the two loyalties.

So much for the husband's gifts. It is worth noting that no equivalent returns are made from the wife's side. The reason, which will be clearer from the analysis I shall presently offer, is that the husband's lineage do not have to divide their rights in him with his wife. But there is one ceremonial act by the wife which is of significance. Soon after the marriage is consummated she comes with her mother and sisters to the husband's home, and with their help cooks a rich and elaborate meal known as 'the great dish' (aduan kese). This is presented to the husband and his kinfolk and friends in sign of the wife's future role as cook and housekeeper.

It is easy to see that Ashanti marriage customs and institutions form a coherent pattern which closely resembles those reported from other African peoples who reckon descent in the matrilineal line. But in order to explain the pattern we must examine the data analytically. In terms of jural isolates, the individual in Ashanti can be regarded as a cluster of rights *in personam* distributed among corporate groups and persons which are not only structurally distinct but may even come into conflict. These rights correspond to distinct interests and fields of social relations symbolized in different ritual concepts and moral rules. They create divided loyalties and represent different kinds of claims which the individual may make on others. Marriage, parenthood, residence rules, domestic organization, and the concept of the personality all fit together in a consistent pattern in the light of this formula.

Marriage, as Professor Radcliffe-Brown has often pointed out, is everywhere a potentially hostile act by each spouse against the kin of the other. In Ashanti this is particularly marked with respect to the wife. For it is in her children that the cleavages between the rights, interests, and loyalties tied to matrilineal kinship and those that arise from paternal filiation, are focussed. There are psychological reasons why an Ashanti woman, like African women in general, never thinks of refusing the responsibilities of adult sexuality, marriage and motherhood. But in accepting them she puts herself and her lineage under an obligation to the father of her children. There are the germs here of an attachment between husband and wife which might endanger lineage solidarity and the rights and claims established by birth and upbringing.

Thus an important feature of Ashanti marriage customs is the affirmation of all existing rights in, and claims on a woman before she enters upon her marital role. This is why the consent of both parents and the lineage elders is required. Again, these rights and claims must be formally admitted by the husband and this is one reason for the gifts. It explains why the gifts are all consumed. The father's sheep, in particular, is not returnable since his rights and claims do not change with changes in his daughter's marital status. By sacrificing the sheep to his soul he accepts the new interests and bonds of marriage and motherhood which will soon come to rival his daughter's attachment to him. This is final whether or not she remains with her husband. The mother's *personal* claim on her daughter's companionship is treated in the same way. Her *jural* rights in her daughter are merged in those of the lineage.

But the crucial issue is the division of rights and claims between the wife's lineage and her husband. The 'tiri nsa' settles that. The fact that it is consumed on the spot and that it resembles the thanking gift which seals a transfer of rights in property, shows, as clearly as the words spoken on the occasion, that it is not a compensation for the loss of a woman of the lineage or a means of bringing about an exchange of a daughter for a wife. It is an agreement about the division of rights in the woman *qua uxor* and *qua filia*. The question of rights *in genetricem* does not arise since these are inalienably vested in the lineage and are subsumed in the woman's obligations as a daughter of the lineage. In a sense it can be said that the 'head rum' creates rather than transfers the husband's rights in his wife but the procedure on divorce shows that there is an element of lineage control over them. This is connected with the right of the lineage to pawn a member – in the case of a woman, virtually to mortgage her to her husband for a loan – for the benefit of the group, and with the corresponding obligations of supporting needy members, materially and ritually, in life and of performing the funerals of all members.

What then are the rights conferred on a husband by the 'head rum' gift? First and foremost, the sole right to the sexual services of his wife, then the right to her domestic services as cook and housekeeper (symbolized in her cooking of the 'great dish'), then to her economic services in his work of earning a living. He does not acquire any right to take his wife away from her maternal kin or to decide where she shall live. He gains no jural rights over her children begotten by him. But the fact that he is their physical father gives him a moral claim, backed only by ritual sanctions, to establish his paternity. Furthermore, these rights entail the obligation of identical return services to his wife, epitomized in his responsibility for any debts she incurs. Debts result from economic activities and after marriage a woman's working power is, in theory, devoted entirely to helping her husband and so contributing to the support of their young children. Debts stand for the antithesis to lineage ties for they cannot arise between lineage kin, who, by definition hold all possessions jointly. The manner and incidence of divorce gives empirical confirmation of the equality of spouses in the marriage relationship.

The rights retained by the lineage are symbolized in the declaration about treasure trove. The maxim applies to men as well as women. Such a find is a stroke of luck. It might happen to anybody, child or adult. It therefore accrues to the benefit of those who have final jural authority over and responsibility for the finder throughout his life. It signifies simply that matrilineal kinship remains absolutely binding on a woman no matter how she disposes of her physiological capacities. The concept of the person as the focus of a cluster of rights and interests which may be distributed among different fields of social relations is clearly brought out in this rule.

One last point. As is well known from Rattray's works, Ashanti have a preference for cross-cousin marriage. Our analysis indicates why. For as I have shown elsewhere (Fortes 1950) cross-cousin marriage is the simplest way of reconciling the divergent and potentially conflicting rights and interests set in action by marriage. Only by stretching the term almost to a point of meaninglessness could it be regarded as a form of exchange. As the Ashanti put it, if a man's sister's son, who is also his heir, marries his daughter, then his property is eventually used to help support her and her children. His obligations to his sister's son are thus reconciled with his affection for his daughter. If a man's son marries his sister's daughter, then his property eventually passes through his sister's son to his son's son, who will very likely be named after him and has the same soul deity as he has, and this is an even better reconciliation of his conflicting obligations and interests. These statements have a metaphorical as well as literal meaning and confirm what I have said about cross-cousin marriage.

The analytical concepts and methods I have illustrated by this brief account of Ashanti marriage customs and institutions are common currency in social anthropology today. They have been most successfully used in the study of social and political organization, economic institutions, and law. The next step must be to find out how we can use them in the study of those more complex and baffling aspects of culture, religion and mythology, art and music, technical knowledge, medicine and rudimentary science.

6

Ritual Festivals and Social
Cohesion in the Hinterland of
the Gold Coast[1]

Social cohesion, or some equivalent concept, has gained an
honourable place in anthropological literature as a labour-saving
device. To an ethnographer working with an Australian horde,
a nomadic Beduin tribe, or even a people with a strong centralized
government, it may seem a self-evident concept; a first law of
social life like Newton's first law of motion, by which everything
else can be explained. In the Northern Territories of the Gold
Coast, however, there is nothing self-evident about it. In the
limited area north of the White Volta and east of 1°W. long. we
find a congeries of peoples speaking different dialects of the
Mossi-Dagomba language family with an ostensibly uniform
culture, but lacking a centralized political organization. There are
no villages in this country, but for miles and miles, continuously,
one mud compound follows on another. There is often nothing
to mark the boundary between one settlement and another, nor
can exact frontiers between dialect areas be established. A short
ethnographic residence in the country shows that a notion of a
fixed and demarcated tribal unit, either as a linguistic grouping
or as a political grouping owing a common allegiance, does not
exist. There seems, in fact, to be no structural unit larger than the
clan-settlement capable of exhibiting cohesion.

The observations presented in this chapter (based on material
collected during a field expedition carried out in 1934–5 under
the auspices of the International Institute of African Languages
and Cultures) refer to the small corner of this area occupied by
the Tale settlements. The Tallensi are well-known to their
neighbours for the ritual festivals which they celebrate. These

[1] Reprinted from the *American Anthropologist*, **38**, No. 4, 1936.

festivals occur between the cessation of the rainy season in September and the commencement of the next rainy season in April. In this paper I shall deal only with a single facet of the festivals, their significance as a mechanism of social cohesion. To appreciate this, some knowledge, however over-simplified, of Tale social and political structure is necessary.

The Tallensi, being anciently settled agriculturalists, are inexorably bound to place. Local group and kinship group tend to be coterminous, hence the social classification of people is primarily in terms of the settlements. Intra-clan social relationships are of a uniform type all over the country. Relations between settlements, that is political relations, however, are of a different order. In keeping with the dominant trend of Tale social life towards decentralization and divergence, rather than centralization and convergence, every clan puts an egocentric interpretation upon its relations with other clans. These interpretations are compounded of prejudice, conventional opinions, traditional attitudes, and personal values; but they conform to a pattern nevertheless. Lacking a central machinery of government and a common allegiance, the Tale settlements have no permanent political relations with one another. Even the Administration has been able to impose only a limited and superficial degree of co-ordination. Traditionally and to this day the heads of clans co-operate in certain contingencies or in certain periodically recurrent ritual situations only. For the rest local autonomy is absolute. The upshot is that the political relations of the Tallensi are a sum of the political relations of each settlement with its neighbours. This means first, that geography is an important factor in their politics, and second, that rigid political frontiers do not exist.

A glance at the political geography of the clans with which the present paper is mainly concerned is therefore essential.

These Tale settlements fall into two major groups. On the one hand are the clans known as the Namoos, on the other the Talis clans. The principal settlement of the Namoos is at Tongo, which was founded, according to the tradition stereotyped throughout Taleland, by Mosur, who fled thither from the Mampuru country in the south, probably over two centuries ago. Mosur was the ancestor of all the people of Tongo and of several settlements colonized from there. Later immigrants from Mampuru established co-clans of Tongo.

The Talis (i.e. the 'real Tallensi') can be subdivided into two groups, those living north of Tongo, the principal settlement of whom is at Baari; and those living on and around the Tong Hills, whom I call the Hill Talis. The Gbizug lineage is in a special category, as we shall see. Talis traditions attribute diverse origins to their clans. Four clans, including Baari, Gbizug, and Wakyi, claim primacy of rank in virtue of the fact that their ancestors emerged from the earth or descended from heaven, and were there when Mosur arrived. The other clans are offshoots of these, or of immigrant origin. The Talis have an ingenious system of clan concatenation, based on a fiction of kinship. Broadly speaking, every lineage of every clan is linked to a lineage of every neighbouring Talis clan by a fiction of half-brotherhood which binds them to reciprocity of certain privileges and obligations, prohibits intermarriage, and especially unites them in a common religious cult. The Namoos clans conform in all respects to the classical definition of a clan; the others must be designated thus for brevity's sake and for want of a term which will convey how they deviate from the prototype.

Namoos and Talis have a common cultural idiom, just as they have a single language. Their economic system is uniform and inclusive of both communities; their laws of land tenure, of inheritance and successions, of marriage and legitimacy are identical. The ritual practices and the mystical notions of their religious and magical institutions are the same in form and dynamical character, even if they sometimes differ in content, especially in the domestic cult of the ancestors. This common corpus of social definitions and of pragmatic organization is the inevitable correlate of the conditions of social and economic intercourse normally prevailing between the two communities. They have been intermarrying for generations, and this entails common legal techniques and principles, as well as a single type of domestic organization. They exchange commodities in trade, gifts, and in the discharge of kinship obligations: for every Namoo has numerous cognatic kinsmen among the Talis and the Talis among the Namoos. Kindred come into frequent contact with one another, not only on ritual occasions, such as sacrifices to ancestors and in funeral ceremonies, but also in ordinary social and personal affairs.

Upon this homogenous basis of social and economic relations between individuals and familial groupings is superposed a political structure the essential principle of which is a polar opposition defined and emphasized by the most stringent ritual observances. The head of the Tongo Namoos has the title Naa or Chief. The chieftainship (*naam*) was brought to Tongo by its founder, Mosur, from Mampurugu. It remains the absolute prerogative of his descendants. Every new chief of Tongo buys his rank from an hereditary elector who represents the chief of Mampurugu. The colonies and co-clans of Tongo elect chiefs of lesser grade. But no Talis may hold the title Naa. Their heads of clans and lineages are typically entitled Tendaana (literally 'owner of the land'). The principal Tendaanaships in the land are attained by rights of patrilineal succession. The chief of Tongo claims suzerainty over all the Tale settlements. Actually his executive authority reaches only to his own settlement, while a Tendaana's executive authority hardly extends beyond his own lineage. Balancing the chiefs' political claims, the Tendaanas assert their precedence in virtue of their 'ownership of the land'. Moreover, each of the principal Tendaanas claims precedence over all the others, such is the degree of local autonomy. Normally an occasion never arises where these rival claims can be pitted against one another. They are vaunted in the presence of the ethnologist, or in the ritual and domestic assemblies of kinsfolk.

Chief and Tendaana, Namoos and Talis, are further separated by a barrier of taboos, many of them symbolic. The chief may not tread upon the earth with his bare foot; he may not pluck a blade of grass or engage in agriculture, and so on. Tendaanas observe none of these taboos. They again may not wear any cloth garment, but only skins, for a white gown of cloth is put upon the chief at his investiture. Cloth garments, horses, guns, are traditionally tabooed to all Talis, since they were characteristic of the Namoos. Chief and Tendaana regard these taboos as of the very substance of their respective offices, as moral obligations to their respective communities. So also property lost on the land or stray animals must be handed over to a Tendaana on pain of supernaturally inflicted death. But stray cows, dogs, or vagrant humans go to the chief. Most important of all, the religious practices of the Talis are dominated by the cult of the earth shrines (*tɔŋbana*) – sacred groves, streams, pools, etc. – to which the Namoos as a group have no access.

The polarity thus institutionally registered is paralleled by standardized political attitudes. In the sporadic fights of pre-European days Talis and Namoos were traditional enemies. This is not only recollected in tales but vividly expressed in the military pantomime which accompanies every funeral ceremony. Talis at their funerals fling taunts and challenges at the Namoos, and the latter retaliate at their funerals. When his patriotism is aroused, a member of one group will speak of the other group superciliously or derogatorily, forgetting for the moment, perhaps, that they are his own mother's kinsmen.

Between Chief and Tendaana, between Namoos and Talis, there are barriers, but not a gulf, an equilibrium and not an irreconcilable disjunction. The fulcrum of this equilibrium is the Tendaana of the Gbizug lineage. Geographically placed between the Namoos and the Talis, he is the ritual and political mediator between them. The principal Tendaana of Baari can directly approach the Chief of Tongo; but the Tendaanas of the Hill Talis or of non-Tale clans can only approach the Chief through the Gbizug Tendaana. In the old days it was he who made peace between the traditional enemies after a fight. A ritual and social equilibrium of high tension exists between Gbizug and Tongo, symbolized in the peculiar relations of the Tendaana and the Chief. Tradition, distorted by each side to its own aggrandisement, relates that it was the first Gbizug Tendaana who received and gave land to settle on to the ancestor of the Tongo Namoos. The final and most solemn of the rites by which a new chief of Tongo is inducted, rites which will be referred to again, must be carried out by the Gbizug Tendaana. Upon him rests the responsibility and the privilege of making the sacrifices for the most sacrosanct fetish of the Tongo Chieftainship, that which safeguards the life and well-being of the Chief and thus the prosperity of the land. Though it is housed in his compound, neither the Chief nor any of his clansmen may set eyes on it; only the Gbizug Tendaana and his lineage may. The Chief speaks of it with an awe which is almost terror. The balance is adjusted by the Chief's rainmaking powers. Only Namoos may own rain medicine, which was brought to Taleland by their ancestor. When drought is prolonged, all the Tendaanas, led by the Gbizug Tendaana, formally call upon and implore the Chief to see to it that rain falls. So great is his rainmaking power that rain will fall at once if he merely declares, in the name of his ancestors, that their wishes will be satisfied. Such, in fact, proved to be the case in 1934.

Again, this ritual polarity is paralleled in social and personal attitudes. The Chief and the Gbizug Tendaana, in private conversation, comment with a mixture of pride, scorn, and fear each upon his opposite number.

Among the most striking mechanisms of this equilibrium are the ritual festivals. The various settlements are always classified in terms of them, and the natives never tire of dilating enthusiastically upon them.

Ritual and dance are the two components of every festival; the former usually esoteric, the latter public though exclusive. During the rainy season, which lasts roughly from April to September, ritual and group activities are almost completely ousted by agriculture. In July-August the early millet is harvested, but the major crops, the guinea corn and late millet, still stand.[1] The Tallensi have a lunar calendar, and the last month of the rains is called the 'Moon of Waters' (Kuom ŋmarig). An ethnologist cannot, alas, be in two places at the same time. I was living among the Namoos in September, 1934, and shall therefore describe the events of this and the next month from their point of view.

Weeks in advance the approaching festivals are the dominant theme of conversation for man, woman, and child. The ceremonial cycle is inaugurated by the Baari Tendaana, who ritually 'throws away the water', i.e. abolishes the rains on the first day of the new month. This is announced by a burst of hallooing, which commences at Baari and sweeps across the country, the signal that the festivals have come round again.

That night the people of Baari begin their celebrations. At Tongo, on the next day, the youths and maidens are putting finishing touches to their festive costumes. On the fourth day of the moon a deputation of elders waits on the Chief, to remind him of his traditional obligation to summon a diviner on the following day.

From this divination emerges, first, whom the ancestor spirits have selected to bring the sacred Gingauŋ drum, after which the Namoos' festival is named, out of the Chief's court chamber.

---

[1] The calendrical incidence of the ritual festivals and their relation to the annual productive cycle are more fully set out in Fortes, M. and S. L. 1936.

There are two Gingauŋ drums: a small one, the sacred one brought by their ancestor Mosur, and a large one, subsequently manufactured merely to add liveliness to the dance. Second, the seance establishes what sacrifices have to be made upon the shrines of the original ancestors of Tongo. In this seance the Chief and his elders, the representatives, metaphorically speaking, of the clan conscience, show how vividly the hallowing of the tradition is felt by them. The sacrifices have a double intention; to thank the ancestors for a successful harvest of early millet and for the standing crop of guinea corn, and to avert quarrels and disputes which might occur when great concourses of people flock to the dance. It is for fear of quarrels which might lead to bloodshed and thus to unknown supernatural calamities that only a youth authorized by the ancestor spirits may inaugurate the dance. The divination, as befits the joyful occasion, is nevertheless cheerful.

The sacrifices are carried out immediately, and ritual then sinks into abeyance for a time. The nightly dance occupies the centre of social interest, until the moon begins to wane. For a night or two the children have almost complete possession of the dedicated dancing field, practising their awkward steps. With the waxing moon adults throng to the dance, until in the end even young mothers come out carrying sleepy infants, and the drummers cannot be seen in the centre of the great mass of people.

This first phase is aptly known as the Gingauŋdeema, 'play Gingauŋ'. It is monopolized by the younger adults and adolescents, who try out the dancing songs. But the season gets its name mainly from the sexual effervescence associated with it. Women are the cause of the disputes which threaten the peace of the dance. Not a day passes without an elopement, planned on the dancing ground. Unfortunately, few of these marriages survive the festive period; and that, perhaps explains why there is a taboo against carrying out any of the legal formalities of marriage during this month. For Baari and Tongo, the Moon of Waters is a dangerous month, in spite of the festivity, because of the human passions aroused, as is evident from the fact that it is a grave sin to shed blood on the earth, even from a scratch or cut. Strangers, and in particular Talis, flock to watch the dance, and of course to find lovers. But only Namoos may enter the dance.

A fortnight's lull ensues during which the guinea-corn is harvested. Then comes the Moon of Daa, so named from the final phase of the festival, and the 'real Gingauŋ', Gingauŋ mɛŋa. But I must resist the temptation to enter into an account of the dynamics of the dance itself. It is the first Gingauŋ incomparably enhanced. The older men, expert dancers, are in charge, though the younger ones and even women and children have full freedom to join.

A point of psychological interest is the choral refrain which provides the organizing slogan for the dance. It consists of a verse taunting a member of a co-clan of Tongo with some moral failing. To taunt the Talis thus would be tantamount to declaring war.

The festival reaches a climax in the second week of the moon, when it squeezes out every other social activity. In the dance this is marked by a striking bit of pantomime which symbolizes the pride of the Namoos in their chieftainship and chiefly ancestors. Suddenly, long after midnight, when the moon is at its highest and the dance at a pitch of intensity, a dozen solemn figures begin to press slowly towards the centre of the dance. Spectators and dancers make way for them. They stand in rank, shoulder to shoulder, unlike the dancers who form a file. The dancers shuffle, leap and stamp, but they bob gravely from foot to foot, erect, faces set like masks, chanting a low, wordless chant. They wear faded red caps and gorgeous, though sometimes tarnished, gowns – the garb of chiefs – and carry spears. They represent their grandfathers and ancestors who had been chiefs and men of rank. The natives delight both in the dramatic contrast, and in identifying the chiefs whom the mummers represent.

And now ritual returns. On the fifteenth night there is no dancing. On the contrary, not a single person stirs out of doors after dark. For that night all the ancestors of Tongo come to dance Gingauŋ, and he who hears or sees the phantom dance dies at once. Everybody hangs out his or her finest garments for the ancestors to borrow for their dance – spiritually, of course. For they keep a sharp look-out against human thieves, who have been known not to respect the occasion. Next day, towards noon, the Baari Tendaana calls upon the Chief of Tongo, bringing a small pot of consecrated beer. All present must drink of it, and the ancestors too, for a gourd-full is set out to be poured on the grave of Mosur. It is a convivial scene, which instantly becomes serious as the Chief begins to address the Tendaana. This is the gist of his speech: The day has returned for us to meet as our ancestors used to, and to do as they did so that we shall have untroubled sleep, marry new wives, and beget many children. We have had a good harvest. May we all live to see Gɔlib and sow our millet in that moon successfully. May my chiefly ancestors and Mosur permit this and permit us to gain new life so that next year at this time we again celebrate Daa. The Tendaana responds with a similar blessing, saying that Baat Daa (the supreme bɔyar of the Baari people) and Mosur will jointly prosper the land. Then he departs. That day Baari celebrates Daa.

That same evening the large Gingauŋ drum is escorted to Baari by the Gbizug Tendaana and the Wakyi Tendaana, following a ritual path. All Tongo flocks to Baari to dance with great enthusiasm on a traditional spot outside the grove of Baat Daa. Near the dance, but silently aloof, wait the Gbizug and Wakyi Tendaanas. A few hundred yards away, as if indifferent to the dance, young men of Baari assemble, wearing their ritual garb, a goatskin. When the Baari Tendaana and his ritual coadjutors arrive to join them, the Gbizug and Wakyi Tendaanas are summoned. Greetings are exchanged, and then led by the Baari Tendaana, the whole company files away in a most solemn and silent procession, brought up by the youngest members carrying pots of consecrated beer. With utter gravity this procession thrice encircles the dancers who continue to dance as if oblivious to this. Anyone who dares to break through the encircling procession is doomed to a supernatural death. The procession files away again into the sacred grove. There the Tendaanas pour libations of beer to Baat Daa, calling down the very same blessings as had been invoked by the Chief earlier in the day. They drink the beer and send a small pot out to the representative of the Chief, who is among the dancers. Now the Tendaanas' procession emerges, again solemnly encircles the dancers, and files away. At a tree which marks the path to Tongo they stop, and marching round it in a circle, chant a song the burden of which is *Mosur boot kuliga* ('Mosur wants to go home'). Upon this signal, the allusion of which is obvious to all, the dance breaks up, the ritual is over.

That finishes the Gingauŋ dance for a year. The next three days at Tongo are given over to the gayest festivities of the season, the celebration of Daa. This is a New Year's festival, a fitting conclusion to the dancing season. Dressed in their best and gaudiest clothes, the whole of Tongo assembles at the Chief's compound on the day after the rites at Baari. From far and near, Tongo women who have married away return to their homes, many bearing gifts of special food from their husbands. The Chief's house throngs with his people and strangers. His elders bring him presents of guinea fowls and he, sitting in state, has a bullock slaughtered, part of which he shares amongst them. The Chief, it is true, gets the better part of this exchange of gifts, but good form is satisfied. The afternoon is devoted to light-hearted dancing in praise of the Chief.

The Gbizug Tendaana has also celebrated Daa, but very quietly. That evening, however, Chief and Tendaana meet in ceremonial. The scene is the dancing ground. This is a dedicated area which may not be cultivated. For this was the site of the first Gbizug Tendaana's compound when he received Mosur; and it is here that the Gbizug Tendaana receives every newly elected Chief of Tongo. After his investiture the Chief must never set foot in it again.

Towards evening a brother of the Chief arrives at the dancing ground, wearing the red cap and the gown, the insignia of a chief, and followed by boys carrying pots of beer. He takes his seat on the flat rock on which every newly elected chief sits down to await the Gbizug Tendaana. For an hour or two he sits gravely there, receiving the humble greetings of the men of Tongo. He is Chief for the nonce. Then a messenger arrives to announce the Gbizug Tendaana. The latter, escorted by men of his lineage, arrives in the dark and greets the mock-chief. The beer is divided out among those present, and then the Tendaana retires a dozen paces. In a voice inaudible to the Namoos, he calls upon his ancestor, the first Tendaana, and asks for the blessings of a good harvest, wives, children, and happiness in the coming year. The mock-chief dismisses his people and takes the homeward path, followed at some distance by the Tendaana and his people. As soon as he arrives at the Chief's compound, he and every other person in the compound, including the Chief himself, hides behind closed doorways in the rooms. The Tendaana enters a silent and empty compound. He goes to the secret place where the sacrosanct Chieftainship fetish is kept, pours a libation to it, and asks blessings for the Chief and all the land. When he has finished he enters the Chief's room to offer him New Year greetings, and so departs. Thus is the Chief every year renewed by a symbolic drama re-enacting his first installation.

Two more days of festivity remain, during which all Tongo celebrates the New Year. Great quantities of food are cooked; there is visiting and counter-visiting, especially by affinal and cognatic relatives, and every one congratulates his neighbour on having seen another year. But the major interest of Daa is the holocaust of sacrifices which accompanies it. Fowls, guinea fowls, dogs, sheep, goats are slaughtered by the dozen in sacrifice to ancestors and to medicine fetishes. The roots of the latter are renewed, and thus given a new lease of life. There is more dancing to celebrate this, and another visit of the Baari Tendaana to the Chief to invoke blessings for the coming year. When the moon ends a few days after, the Namoos have finally done with their Harvest and New Year festival. Before that the Chief may not eat of the new guinea corn, though commoners do.

Two points deserve notice in this series of ritual celebrations. First, they express clearly the ambivalent relationships between the Namoos and their neighbours, and the bridging function of the Gbizug Tendaana. It is as if they were joined in a mutual responsibility each for the other, based upon a profound antagonism. Second, the rites evidently recreate and regenerate the religious, the magical, and the traditional bulwarks of the social life of the Namoos.

A week later the scene shifts to the Hill Talis. On the first day of the following moon, Bɔyaraam ŋmarig, the Talis begin their harvest festival of the same name. Despite the difference of name, it is the Daa festival translated into the peculiar and esoteric ritual idiom of the Hill Talis. Namoos and Talis, as the exclusion of the former from the inner mysteries at Baari indicated, constitute two religious sects. The cult of the *bɔyar*, after which the Hill Talis name their festival, dominates the religion of all the Talis, but is known to the Namoos only by hearsay. Its hypertrophied development among the Hill Talis is indeed completely hidden from their own sect-brethren at Baari and Gbizug. A *bɔyar* may be a grove of trees, a cave in the hillside, or a small natural enclosure made by trees and boulders. It is truly sacred, for it may be entered only for ritual purposes, and in the company of all the ritual officiants. A *bɔyar's* communicants generally comprise a group of linked lineages or of several clans, each of which appoints one officiant. All the ritual of the *bɔyar* cult is built up on the notion of sacrifice, and every officiant or his representative must be present at every sacrifice. It is an axiom that a quarrel or dispute between communicants must be healed at once, by rites of reconciliation at the *bɔyar*. The *bɔyar* is the pinnacle of their ancestor cult, for it is there that all the ancestors of the communicants reside. It has tremendous magical power, amoral, like all Tale magic, and therefore applicable both for beneficent and maleficent ends. Thus its potency as fertility magic is widely renowned, and it can be invoked to bring death and destruction upon an enemy, by its communicants.

This, and much more, I learnt with very great difficulty, only after I had been initiated. For many months my enquiries were fruitless. The Namoos know the dates of Bɔyaraam, and the dances. Of the rituals they had, despite centuries of intimate contact, absolutely no knowledge. Indeed they refuse to hear of it, for they regard the cults of the Talis with fear and aversion. The Talis, on the other hand, never ceased to warn me against betraying what I had seen to my Namoos friends. It is, for them, the sin against the Holy Ghost punishable by instantaneous death, inflicted by the *bɔyar*. This, the most absolute cleavage between the Namoos and the Hill Talis, depends upon the fact that the latter initiate their sons into their cult. Nothing corresponding to initiation exists among the Namoos or even Gbizug or Baari. The Bɔyaraam ritual is at the same time communion, initiation, and offering of first fruits.

The various *bɔyar* congregations commence the ritual of Bɔyaraam on successive days, starting from the first day of the moon. The order of entry is fixed by tradition, and claims to precedence in rank are based in part on this. But preliminary sacrifices, consultations of diviners, collection of the chickens and guinea fowls which will be required, and finally ritual preparation of beer, usher in the festival. From the 'Moon of Waters' on, every head of a joint family must eat out of special vessels, lest his food or water be contaminated by the newly harvested guinea corn flour, ground beans, or ocra. None of these foods is prohibited to women and younger members of the family. The taboo depends not upon some notions of the magical danger of eating unconsecrated first fruits, but upon the relationship of the communicants to the *bɔyar*, who has the prerogative of tasting them first.

I cannot describe the esoteric rites of Bɔyaraam in detail. The main events take place during one night and the next day. The initiand youths, selected simply by the fact that their fathers can afford the numerous fowls and guinea fowls and the beer necessary for the rites, are pounced upon at night, undressed, and segregated throughout the rites. The communicants wear skins, the ritual garb of the Talis. Women and uninitiated boys are rigorously excluded from earshot. Each elder brings fowls, guinea fowls, guinea corn flour, and cooked ground beans, and sacrifices are made in great numbers both at the houses of the Tendaanas and inside the *bɔyar*. The ancestors are called upon and the *bɔyar* is invoked. 'Your day has come round again. We are about to eat the new guinea corn. Hence we bring you this offering of guinea corn and beans and these, your bɔyaraam chickens. Grant us untroubled sleeping. May our wives bear children, and may we sow early millet successfully and meet again next year.' This is the theme of the sacrificial invocations. Every chicken thus slaughtered is observed with concentrated anxiety. It must come to rest on its back, else the donor has sinned in some way and must at once go to a diviner to discover how. For the initiands long life, success in marriage and in farming, and safety from enemies is begged. It is a very serious thing if an initiand's fowl is rejected by the ancestors; and all proceedings are suspended until the sin is traced and atonement made.

The ritual is sensationally dramatic, and entirely different in style and emphasis from Namoos ritual. There is a constant interweaving of the three motifs, of communion, initiation, and first fruit offerings. The tone fluctuates between intense and almost terrified ritual participation and casual conviviality, when the beer is being divided.

The *bɔyar* is first entered mysteriously in pitch darkness; but the 'high spot' is what might be called the administration of the sacrament to the initiands next day. They have been treated throughout with veiled contumely, like infants and outcasts, seated on the bare ground, their backs to the ceremonial. When the sacrifices are over and the beer consumed, the initiands are brusquely dragged up, each in turn, to the altar of the *bɔyar*. Kneeling before it, quivering with fear, his eyes blindfolded by the hands of one officiant, his hands clasping a chicken whose neck he saws off on a rough stone with the help of another officiant, he is made to swear repeatedly 'If I tell anybody may I die.' This then is the secret of the impregnability of the Talis *bɔyar* cult both to the Namoos and to the ethnologist.

Before they leave, the meat and flour of all the sacrifices are divided out according to strict rules of prerogative; the initiands being fed with a tiny piece of meat and flour dumpling each. The role of each participant kinship group is defined in terms of the officiant it elects, the contributions it makes to the sacrifices, and the share of meat, beer, and flour it receives. The natives stress this organizational balance more jealously than details of ritual; for it is the structural framework which preserves the ritual collaboration. Finally, all the communicants, including those newly initiated, are daubed on brows, arms, and knees with consecrated red mud, the concrete emblem of the beneficent magic they are carrying home.

Bɔyaraam, like Daa, is the occasion when married daughters return home with gifts of food and guinea fowls from their husbands, and men who have migrated to distant parts try to return for the ceremonies. Dancing follows after the ritual day. But as each settlement dances by itself, the Bɔyaraam dancing is neither so impressive nor so great a focus of social interest as Gingauŋ. This time the Namoos are onlookers, for they may not enter the Tallis dances.

The Bɔyaraam festival of the Hill Talis thus appears to be homologous with the Gingauŋ and Daa of the Namoos, and the Daa of Baari, in its calendrical reference and its functional context. For Chief, Tendaanas, and elders, the fathers of the community who bear the greatest burden of responsibility for its welfare, the ritual is the dominant theme. For those of lesser social responsibility, the dancing and festivity, the fellowship of jubilation, are of major value. For everybody, these are festivals of reunion in which family and wider agnatic connections, the unique fact for the individual of having been born into a certain family and clan, receive special emphasis.

But what must chiefly interest us, in this paper, is to observe how these festivals differentiate and bind the Namoos and the Talis. On the one hand it seems as if each community consolidates itself socially and morally by the very act of repulsing its neighbour. Behind its stone wall of exclusiveness, even to the date of its festival, each community celebrates its release from the hazards of the past year and especially of the food-growing season, and fortifies itself by magical and religious techniques for another year. Nothing could more strikingly demonstrate the factors of dichotomy and antagonism in the polarity which we saw previously to be fundamental to Tale political structure than the difference between the rites of the Hill Talis and those of the Namoos. It is true that every Tale ceremonial activity expresses the exclusiveness of the group performing it; but in these festivals the expression of difference has a political validity because it is reciprocal. The difference is published in the dance and dramatized in esoteric ritual.

But Namoos and Talis are culturally equivalent communities, dwelling in close juxtaposition, having intimate economic and social relationships with each other. This, it seems to me, sets a limit to the degree of antagonistic differentiation tolerable; and thus the other factor in the polarity of Tale society comes into existence – the equilibrium between the two communities. The very opposition, as it were, engenders dependence. This aspect of Namoos-Talis relations comes out clearly in the ceremony at Baari, in the visits of the Baari Tendaana to the Chief, and in the ritual re-enacting of the Chief's induction at Tongo. It is notable how these rituals vividly insulate each group from the other, while at the same time uniting them in common responsibility for the welfare of the country.

The same compulsion to co-operate in the magical safeguarding of the land and the people obtains between the Namoos and the Hill Talis. It is expressed in the Gɔlib festival, which unfortunately space does not permit me to describe fully. The purpose of this festival, which occurs in the last month of the dry season, is to ensure a successful sowing and harvesting of early millet and to call down the blessing of fertility in general. In it all the clans of the Hill Talis collaborate ritually and unite to dance together. In their ritual collaboration the distribution of offices and sacred apparatus among the several clans is so remarkably equal that each one of them can claim with ostensibly full justification to be more important than the others. It so happens that a bitter personal feud is at present raging between one of these officiants, on the one hand, and the Chief of Tongo and another of the Gɔlib officiants, on the other. But none of them could escape his obligation to co-operate with his enemy in the appropriate ceremonial situations. Here, perhaps, is a clue to the problem of why ritual forms and religious situations should be the institutional media selected for the expression of the equilibrium in Tale society. They have a compulsive power which a pragmatic institution orientated to the demands of the objective world could never have.

Golib is a dangerous month, during which both Namoos and Tallis are subject to numerous taboos, mainly aimed at averting quarrels and strife at the dances and in everyday relations between the groups. It is forbidden to shed blood on the earth, lest supernatural death overtake one, to take a wife, to carry out the final ceremonies connected with delayed funeral rites, to wear cloth garments in any of the Hill Talis settlements, and so forth. Marriages are the cause of most inter-group litigation, and funeral ceremonies represent social situations in which large numbers of people from different settlements congregate and often give rise to disputes. Throughout, the Namoos are spectators of the dancing and excluded from the ritual. But on two occasions they meet the Hill Talis in ceremonial interaction. In the inaugural rites the Hill Talis, led by their Tendaanas, march to all their sacred groves in succession blowing sacred whistles in honour of them. When they reach the sacred groves which lie in Tongo, every Namoo man, woman, and child hides indoors. A Namoo who sees these rites of the Hill Talis risks the destruction of his entire home and family by the outraged sacred groves. Once again, therefore, are the Namoos and Talis mutually insulated or segregated by the symbolism of unchallengeable ritual, the living affirmation of tradition. But there is a ritual reintegration of the groups. The final rite of Golib requires the co-operation of the Chief of Tongo. Again led by their Tendaanas, the Hill Talis come down to the Chief's house to dance; and there Tendaanas and Chief join together to beg for blessings upon the land, for the safety of the people and the abundance of crops.

To revert to the harvest festivals, the magical and religious value of the ceremonial to the group performing it must not blind us to its socially integrative functions. The former is mainly determined by the seasonal context of the festivals and is evidenced pre-eminently in commonplace techniques, such as the sacrifice. The latter, which is determined primarily by the context of the social structure, is evidenced in the ceremonies peculiar to the occasion. These differ completely in content, as between Namoos and Talis, a fact which the natives would explain by citing the historically different origins of the two groups. This suggests the principle common to all the integrative ceremonies. It is the dramatization of the nodal, hence traditionally sanctioned, relationship in the social structure of each group – the Chieftainship at Tongo, the right of access to the bɔyar among the Talis. This fundamental bond is resuscitated in both the symbolical and the direct phases of the ceremonies. The Chief and the Gbizug or Baari Tendaana thus reimpose on each other those responsibilities which bind them to a common task; and, by the very same rites, the Chief and his people redefine each the place of the other in a social system the essence of which is the Chieftainship. The Hill Talis, again, re-establish the responsibility of each initiated man for the preservation of their traditional cult, by imposing it upon the next generation, and thus reaffirm the loyalty of each to the other, and of sons to fathers.

To sum up, social cohesion as I understand this concept, both within and between the major communities which constitute Tale society, is no ultimate attribute of that society, but is achieved by specific social mechanisms such as those I have described. To me its most significant feature appears to be the fact that this integration is engendered as an equilibrium between opposed groups, over-riding, it seems, the tendencies to conflict inherent in the system. It is worthy of note that this principle of a balance of powers is exploited also within each community to overcome the competitive autonomy of its constituent groupings.

Whether only a single general process is involved, accommodation to an immigrant group, is beyond my power to determine. One could presumably envisage the entire pattern of ritual and political values as a 'reaction' to the 'foreign body' represented by the Namoos clans. It seems not unlikely also that the equilibrium thus achieved is itself the barrier that prevents the diffusion of the esoteric rites of one community to the other.

7

Pietas in Ancestor Worship[1]

I

The Henry Myers lecture is meant to be addressed to a mixed audience of anthropologists and non-anthropologists of varied interests. This makes it a tempting opportunity for stepping off the straight and narrow path of professional specialism to wander in the green pastures of speculation from which one normally averts one's eyes. If I have been rash in yielding to this temptation, I trust that this is not an act of impiety towards the founder of this lecture.

When I began my field work among the Tallensi in 1934, the controversy aroused by Malinowski's assault on the Freudian hypothesis of the Oedipus complex was still simmering. Though he himself was in the phase of behaviouristic revulsion against psychoanalysis, his earlier views remained influential both as a challenge to psychoanalytical theory and as a stimulus to anthropologists in the field. Just what sort of field investigation of the problems in dispute could feasibly be attempted by an anthropologist, was the subject of lively discussion. The main theoretical issues had been elucidated with characteristic impartiality and clarity by Seligman, in his Huxley Lecture (1932). But all that was certain was that the primary social field within which the oedipal drama might be expected to manifest itself in custom and behaviour was that of family and kinship. The tricky question of what inferences could legitimately be made from overt customary behaviour to the hidden motives and fantasies identified by psychoanalysis, was left unresolved, as indeed it still is to a great extent (*cf.* Leach 1958). However, a distinctive orientation towards both field work and theory emerged as the ethnographic monographs and studies of the thirties and forties testify. As it shaped itself for me there were three basic rules. The first, learnt from Malinowski, was that a custom, or body of custom, what-

[1] The Henry Myers Lecture, 1960; reprinted from the *J.R.A.I.*, **91,** Part 2, 1961

ever its historical source may have been, is meaningful in the contemporary social life of a people, and that the anthropologist's essential task is to investigate this fact. The second, learnt from Radcliffe-Brown, was that custom is embedded in social structure and is significant of social relations. The third, due to the prevailing climate of psychological thought, was that custom is the socially tolerable expression of motives, feelings and dispositions that are not always acknowledgeable and may include potentially disruptive as well as constructive elements.

II

Kinship and ancestor cult are so prominent in the household and neighbourhood arrangements, the economic pursuits, and the routine of social relations among the Tallensi, that I was obliged to make myself adept in these matters from the outset of my field work. I arrived in the middle of the dry season. It is the time of the year when funerals are celebrated, both because the weather permits and because there is grain for beer and leisure from farming. For the same reason, it is also the preferred season for communal ceremonies and for many major domestic rituals.

Thus far from being in a position to establish my good faith by showing an interest in such neutral topics as string figures (which have no significance except as an amusement for children), material culture, or crops and markets, I was flung straight into divination, funeral ceremonies, domestic sacrifices, and the Harvest and Sowing Festivals. And it quickly became apparent that no understanding of these ritual and ceremonial activities was possible without a thorough knowledge of the kinship, family, and descent structure. For the Tallensi, like most African peoples with a highly developed system of ancestor worship, patently associated with descent groups and institutions, fit very well the paradigm of the religious community, in what he spoke of as 'early stages', sketched with such masterly insight by W. Robertson Smith (1927). I refer to his observation (p. 54) that 'it is not with a vague fear of unknown powers, but with a loving reverence for known gods who are knit to their worshippers by strong bonds of kinship' that religion begins. He was struck by the parental characteristics of early Semitic divinities and connected this with the composition of the congregation of worshippers as invariably a 'circle of kin' whose greatest kinsman was the worshipped god. 'The indissoluble bond that united men to their god', he concludes, 'is the same bond of blood fellowship which ... is the one binding link between man and man, and the one sacred principle of moral obligation' (p. 53). And particularly worthy of recollection, for its bearing on my theme today, is his comment (p. 58) that 'the feelings called forth when the deity was conceived as a father were on the whole of an austerer kind' than those directed to a maternal deity because of the father's claim to be 'honoured and served by his son'.

Robertson Smith was not the only scholar of his generation who perceived the connection between the institutions of kinship on the one hand, and religious beliefs and practices on the other. He was indeed anticipated by a quarter of a century by that other inspired precursor of our current ideas, Fustel de Coulanges, to whom I am specially indebted. But for him the linkage was to all intents the other way round. Where Robertson Smith supposed parenthood and kinship to underly the worship of their gods by the Semites, Fustel argued (1864, bk. ii, ch. v) that it was the ancestral cult of the Romans which imposed agnatic kinship. 'The source of kinship', he says, 'was not the material fact of birth; it was the religious cult;'[1] and he goes on (ch. VII) to demonstrate brilliantly how succession and inheritance are interlaced with the domestic ancestor cult. I quote: 'Man dies but the cult goes on. ... While the domestic religion continues, the law of property must continue with it,' and further, with regard to the law of succession, 'since the domestic religion is hereditary ... from male to male, property is so too ... what makes the son the heir is not the personal wish of his father ... the son inherits as of full right ... the continuation of the property, as of the cult, is an obligation for him as much as a right. Whether he desires it or not it falls to him.' The essential point, by his reasoning, was that in early Greek and Roman Law descent in the male line exclusively determined the right to inherit and succeed to a father's property and status but it was primarily a religious relationship. Hence a son who had been excluded from the paternal cult by emancipation was also cut off from his inheritance, whereas a complete stranger who was made a member

[1] This is my translation of his statement: 'Le principe de la parenté n'était pas l'acte matériel de la naissance, c'était le culte.'

of the family cult by adoption thus became a son entitled to inherit both the worship and the property.

Robertson Smith was not immune from the fallacies of his day and has been justifiably criticized for this,[1] and Fustel, I understand, is considered by some Classical scholars to have subordinated scholarship unduly to conjecture. Be this as it may, we cannot but admire their perspicacity in directing attention to the social matrix of the type of religious institutions they were concerned with. For at that time the orthodox approach to early religions was by way of their manifest content of belief. From their pinnacle of intellectual rectitude, most scholars saw no further than the false logic, the erroneous cosmology, and the emotionally distorted superstitions which their pre-conceived theories revealed in non-Christian religions. This was the school of thought whose first concern was with what Robertson Smith designated as the 'nature of the Gods' and with which he contrasted his own procedure (*cf.* p. 8). This is the tradition of Tylor, Frazer, and Marett, and the host of their followers, amplifiers and expositors too numerous to list (and mostly now quite obsolete). And this, fundamentally, purified of its grosser bias, is the tradition of Malinowski and of Lévy-Bruhl, as well as such famous ethnographers of Africa as Rattray and Junod, Westermann and Edwin Smith.

I lay no claim to having been aware of the bearing of Robertson Smith's and Fustel's theories on the religious institutions of the Tallensi when I studied them in the field. That came much later. It was simply that ancestor worship was too conspicuous to be missed and that the framework of genealogical bonds and divisions was an aspect of ritual to which both participants and commentators freely drew attention. But given the general orientation I have described, what started me thinking about the crucial factors of ancestor worship was the casual observation recorded in the book which originally aroused my interest in the Tallensi.

In 1932 there appeared the first systematic ethnographical survey of the tribes of Northern Ghana, R. S. Rattray's *Tribes of The Ashanti Hinterland*. It is, in fact, a somewhat disconnected compilation of Rattray's own observations and informants' texts. But with his uncanny knack for field enquiry, in following

---

[1] Stanley A. Cook's Introduction and Additional Notes to Robertson Smith (1927), Third Edition, deal with this at length.

up some of the kinship customs of the Nankanse, who are
neighbours of the Tallensi and differ little from them in language
and culture, Rattray (1932, p. 263) discovered a rule which he
reports in these words:

Among the Nankanse, as also among many other tribes, it is forbidden for the first-born (male and female) to make use of any personal
property belonging to the parents, e.g. to touch a father's weapons,
put on his cap or skin covering, to look into his grain store or into his
tapo, leather bag, or in the case of the female, to pry into her mother's
kumpio. 'Parents do not like their first-born and it is unlucky to live
with them.' I think [comments Rattray] the idea is that they are
waiting, as we would say, 'to step into the dead man's shoes'.

That parents and children are often opposed and even antagonistic to one another is widely acknowledged. It is a common
enough theme of European novels and plays. Anthropologists
have long been familiar with the parallels in primitive society.
But its cardinal importance in social life was only beginning to
be understood in 1932, partly through coming into the limelight
of psychoanalysis[1] but more particularly through the kinship
studies of Malinowski and Radcliffe-Brown. Radcliffe-Brown's
revolutionary paper on the Mother's Brother (1924) had made
us realize the significance of respect and avoidance customs as
expressions of the authority held by fathers over children in a
patrilineal family structure, and Malinowski (1927) had revealed
the conflicts that go on under the surface of matrilineal kinship
norms (*cf.* Fortes 1957). The Nankanse custom seemed to betray
outright hostility between parent and child of the same sex,
linked to open admission of the wish for the parent's death. It
was curious also in singling out the first-born. No anthropologist
alert to the current controversies concerning kinship and family
structure could fail to be intrigued.

III

The avoidances observed by first-born children towards their
parent of like sex were quickly and frequently brought to my
notice among the Tallensi. As I have described elsewhere (1945,

[1] J. C. Flugel's pioneer book, *The Psychoanalytic Study of the Family* (1921),
deserves grateful acknowledgement for its influence in bringing together psychoanalytic theory and ethnological research in kinship studies.

1949, 1959) they are not only a matter of common knowledge but have a critical significance for Tale social structure both within and beyond the domestic domain in which they are primarily operative. The personal avoidances of the Namoo first-born are public moral obligations in that adherence to them is a symbol of alinement by clanship. Not surprisingly, therefore, first-borns often spoke of their situation with a tone of pride, though it might seem, to the outsider, to be irrationally burdensome and fraught with the humiliation of rejection. Of course, first-borns are inured from babyhood to the disabilities of their status and these are ritual injunctions stamped with absolute inviolability from the outset. But what a first-born is inured to looks, to the outsider, more like a deprivation loaded with threat. From early childhood he must not eat from the same dish as his father lest his finger scratch his father's hand. Tallensi say that if this happened it would cause misfortune, possibly even death, to the father. But he sees his younger brothers share their father's dish with impunity. If they scratch his hand by chance no harm ensues. And it is the same with other observances – the prohibition against wearing his father's clothes, using his bow, and looking into his granary. Yet first-born sons do not speak with resentment of their situation. They accept it with equanimity, with good grace, often, as I have said, with a kind of pride. It is, quite simply, from their point of view, a rule of life, in their language, an ancestral taboo (*kyiher*); but not an arbitrary and irrational one.

This is of great importance. It is accounted for in terms of rational interpretation of the social and psychological relationships of fathers and sons – an interpretation which makes sense equally from the point of view of Tale social structure and Tale values and from that of the anthropological theory of kinship. To see how apposite it is, one has to remember that Tale fathers are devoted to their children and are not denied by custom the freedom to show affection and familiarity towards their sons. The picture given by Professor Carstairs (1957) of the attitude of fathers to their children, and especially of the relationships between fathers and sons, among the Hindus of Rajasthan would horrify Tallensi; they would not even agree unreservedly with the maxim that a 'man must always defer unhesitatingly to his father's word', though they insist that fathers must be accorded respect and obedience. Carstairs makes much of the aloofness, the lack of spontaneous warmth and intimacy, in the relations of father and son, and connects this with the strict obligations that exist for both sides. The Tallensi are not so severe. They do not idealize the father, like these Hindus, as a feared and remote model of austerity and self-control in whose presence everything associated with pleasure, levity, and most of all sexual life is forbidden. For them, too, the ultimate disaster, beside which death itself is insignificant, is to die without a son to perform one's funeral ceremonies and continue one's descent line. But having no notion of an after life corresponding to hell and *nirvana*, they would not properly comprehend the Sanskrit maxim quoted by Carstairs (1957, p. 222) that 'a son is he who rescues a man from hell' and assures his attainment of *nirvana*. As we shall see, for Tallensi, to have a son is to ensure one's own ancestorhood, and that is all the immortality one aspires to. With their nebulous ideas of the mode of existence of ancestors, Tallensi do not have beliefs that it can be influenced by the conduct of offspring and descendants.

I have cited Carstairs's observations in order to bring out the relatively rational attitude and amicable compromise achieved by Tale custom, at any rate in its manifest and public aspect, in handling the tensions in the relationship of father and son, as focused upon the first-born son. As I have shown elsewhere (1949), this fits in with the lineage system, the domestic organization, and the widely ramifying web of kinship that together form the basis of Tale social structure. As in all societies in which patrilineal descent is the key principle of the social structure, the relationship of fathers and sons is the nuclear element of the whole social system. Opposition and interdependence, to use rather non-committal terms, are mingled in their customary conduct to one another. But what is of particular interest for my present theme is that Tallensi recognize quite frankly and even with some irony that the opposition between fathers and sons springs from their rivalry. Moreover, they regard this rivalry as inherent in the very nature of the relationship. If they had the word they would say that it was instinctive. In particular, they perceive that the prohibitions followed by first-born sons not only give customary expression and legitimacy to this fact but serve as a means of canalizing and dealing with the potential dangers they see in it.

They are also well aware of the economic, jural, and moral factors in the situation. (Fortes 1949; 1959).

It will help to clothe this bald summary with verisimilitude and to lead to the next step in my argument if I stop to give you a typical instance from my field records. Saa of Kpata'ar, a man of about forty-two, was showing me his home farm one day and explained that until five years before he had lived and farmed with his father.

'Then,' he said, 'my father told me to come and build a house for myself and my wives and children on this land where his grandfather had formerly lived and to farm and provide for my wives and children by myself. Now he lives in his house and my younger brother, and his own younger brother live and farm with him. [Then, with a humorous glint in his eyes and half smiling, he went on] You see, I am my father's eldest son. In our country, for all of us, whether we are Talis or Namoos, this is our taboo; when your eldest son reaches manhood he either goes out on his own or cuts his own gateway in his father's house. If my father and I were to abide together it would not be good. I would harm us. It would harm me and would that not be injurious also to my father?'

Now I knew Saa's old father quite well and had often seen him at his father's house assisting and advising in family and lineage affairs. Their public intercourse seemed to be as friendly and their mutual loyalty as staunch as was overtly the case with most elderly fathers and their mature sons. I therefore pressed Saa further, asking how he and his father could conceivably harm each other merely by living in the same household and farming together. He replied in the same matter-of-fact manner in which he had earlier given me some details of his farm work.

'It's like this', he said, 'if we abide together our Destinies wrestle with each other. My Destiny struggles that he shall not live and his Destiny strives that I do not live. Don't you see, there sits my father and he has his ancestor shrines; if he were to die today it would be I who would own them. Thus it is that my Destiny strives for him to die so that I can take over his shrines to add to my Destiny and his Destiny strives for me to die so that my father can keep his shrine to sacrifice to them.' [He spoke as if he were describing the action of external forces that had nothing to do with his own will and desires. I remarked that this suggested a standing enmity between his father and himself. He responded in a more personal, but still philosophical tone of voice.]

'Indeed', he mused, 'we don't like an oldest son, we care for the youngest son. As for my father, of course I am attached to him. If he were to die today I would have a hard time of it. His younger brothers would take possession of the family home. I am only a minor person. It is because my father is head of the family, on his account, that I have a house of my own and possess this farm. If he were to die his next brother inherits the family property and that includes what I have. Nowadays, if a big sacrifice is performed my father gets his share and he gives some to me; his brother would not do this. Nowadays, I act on my father's behalf in public affairs; when he dies I shall become a nobody. I want him to live.' [I pointed out that he seems to be contradicting himself; to which he replied:] 'When my father and I dwelt together he used not to heed what I said. If there was a dispute he would listen to the others, never to me. Now that he does not see me every day and I have my house and he has his, his soul at length turns towards me. True, I only left him recently. That was because I was not ready for it before. As I farmed with him I was entitled to have him pay the bride-price for any wife I married. When he had provided for me to marry to my satisfaction I was ready to go out on my own. My younger brothers have remained to farm with him. They can't inherit his ancestor shrines when he dies, so their Destinies have no quarrel with his.'

This revealing confession sums up the normal and conventional conception of the relations of men with their sons among the Tallensi. Many such statements, supported by observations of people's attitudes and behaviour in many situations, confirm its accuracy and sincerity. What Saa tells us is that there is a latent antagonism between a man and his eldest son all through life. In the son's youth it does not stand in the way of their daily association and their amicable co-operation in farming and other household concerns. But when the son marries, and in due course becomes a natural father in his own right, responsible for the support and care of wife and children, his further growth in social achievement and personal maturity begins to be felt as a threat to the father. Unconscious antagonism turns into potential strife.

What is at stake is clear enough. It is the status of fatherhood. This is exhibited in the possession of the rights of disposal over family property, but, more significantly, is conferred by attaining custody of the ancestral shrines. But what matters most is that it is a unique status. Given the patrilineal lineage system, there can be only one father-of-the-family, in the sense of the person vested with supreme authority in the family, at any time. And there is only one way in which this status can be attained and that is by succession. But this presupposes the death of the holder (to borrow Dr Goody's valuable concept (Goody 1959) which brings out the natural transiency of such holding). And this is the crux. To safeguard the rightful holder from the competitive aspirations of his rightful heir-apparent is the issue. It is presented as a curiously impersonal issue, as if it were a given fact of human nature. And this fits in with the way it is dealt with by means of the quasi-impersonal imperative of taboo. What is accomplished, in fact, is the segregation of the protagonists from each other in respect of the two primary spheres of paternal authority, the control over property and dependants and the monopoly of the right to officiate in the worship at ancestor shrines. As Saa claims, voicing common sentiments, relationships of goodwill and mutual affection on the personal level are not disturbed. These go back to a father's devoted care for his children during their infancy. And we can see the point of the avoidances imposed at this stage. In the patrilineal and patrilocal joint family system of the Tallensi, fathers have to support, bring up and educate their sons to follow in their footsteps and succeed them. A son cannot be socially and materially segregated in childhood.

The social structure and economy of the Tallensi rules out the possibility of sending a child away to be brought up by maternal kin, as is done among the Dagomba, for instance, nor are the age-villages exploited by the Nyakyusa (Wilson 1951) to segregate successive generations feasible.

How can the son, designated by jural and ritual custom to be his father's successor, be kept submissive to paternal authority in childhood except by excluding him from activities and relationships that smack of sharing in his father's status? How can the son be equipped to perceive and feel his obligatory separation from his father; how can he dutifully make plain that he is not his father's equal and does not covet his father's position? Physical separation being ruled out, the answer is found in the symbolic avoidances described earlier. In the circumstances they have to be more explicit and categorical than the forms of etiquette by means of which respect for parents and elders are shown in some African societies, even, for instance, among the Thonga (Junod 1927, p. 441 *seq.*), where the eldest son's situation is very close to that of the Tallensi. For what is demanded is more than respect though less than the extreme spatial and politico-ritual insulation of filial from parental generation that is found among some Central African peoples. (I have instanced the Nyakyusa, but the custom is widespread in Central Africa.) What is symbolized is that an eldest son must not pretend to equality with his father as economic head of the household (hence the granary taboo), in respect to the rights he has over his wives (hence the taboo on eating with the father, since one of the main duties of a wife is to cook for her husband) in his status of mature manhood, as an independent jural and ritual person, which can, as I shall presently emphasize, only be reached upon the death of the father (hence the taboo on the father's bow and quiver), and lastly, in his capacity as a unique individual (hence the ban on wearing his father's clothes). And these observances must be kept – I can vouch that they are so kept – without destroying the personal warmth and trust that is also an essential component in the relationship of the son with his father.

This will be clearer if I hark back to note that what I said earlier about inheritance and succession was elliptical in one respect. I should have pointed out that a father of a family has two distinct elements of status. He is father of his children by right of begetting, as Tallensi say, and by this token his sons are simply extensions or parts of himself during his life time. They have no jural standing in their own right, even if they are economically self-supporting and live separately. This is a basic norm of Tale social structure. I was vividly confronted with it when a young man employed by me appealed to me with mingled anger and resignation. He had married a girl by elopement and the placation gifts had been accepted. But his father had refused to complete the formalities on the grounds that he could not afford to pay the bride-price cattle. Yet my young friend had saved enough to buy two cows which would be an ample first instalment on the bride-price. Why, I asked, did he not, then, himself hand over the cows to his father-in-law? It would be outrageous, he replied, even if the wife's lineage kin were unscrupulous enough to accept them. You can't pay your wife's bride-price yourself while your own father is alive, not even while one of your father's own brothers is alive. It would be 'setting myself up as my father's equal', he explained. 'We should quarrel, he would curse me and refuse to sacrifice on my behalf; does not one man surpass another in standing?'

His one hope was that I might persuade his father to give in. But note an important corollary. My informant would have been no better placed, jurally, to take a wife even if he already had one wife and children of his own; for a man does not have the jural autonomy to act independently, on his own behalf, even in regard to rights over his own children, until his father dies.

The other side of a father-of-a-family's status is his position as head of the lineage segment which constitutes the core of the family. He arrives at this status, not by having children or by succeeding his father by right of filiation, but by succeeding to it by seniority in the lineage. By this reckoning lineage brothers succeed first and then sons; and of course all brothers who survive can in time succeed. There is not the specificity of filial succession. Thus Tale fatherhood conforms to Maine's (1861, ch. VII) dictum that 'patriarchal power is not only domestic but political'. Taking fatherhood as a status in the politico-jural domain, in which the relationships between holder and prospective heir are modelled on siblingship, we see that there are no avoidances between a man and his prospective lineage successor either of a ritual or of a secular nature. Brothers borrow each other's clothes and may inherit each other's widows. As lineage head a man cannot frustrate his brother or his brother's son in the matter of bride-price as arbitrarily as a father can, nor can he refuse to sacrifice to a common ancestor without grave cause.

We must conclude that the first-born's avoidances are to be understood as referring to his father's strictly paternal and his own strictly filial status in the domestic domain during his father's life time. This is an inescapable nexus; and this explains why Tallensi account for the opposition of father and first-born son, prescribed by custom though it is, by means of spiritual concepts rather than in jural and economic terms. Breach of the taboos would be an affront to the father's soul (*sii*) and Destiny (*yin*). For a man's soul is in his granary and his vitality is in his garments and his weapons because they are covered with the sweat and dirt of his body. This is more vividly brought out among the Talis clans which do not impose avoidances between father and first-born as taboos, though they follow the practice for what they regard as reasons of propriety. Among them a first-born son may be sent by his father to fetch grain from his granary. But when a father dies his bow and quiver and leather pouch are hung up inside his granary by the officiating elders. From that day until the final obsequies, which may not take place for two or three years, the eldest son may not look inside the granary. If he did, he would see his dead father and himself die. A younger son can enter the granary with impunity. He would never see the dead. At the final obsequies the hidden articles are brought out and eventually taken by the eldest son to be deposited in the sacred grove of the lineage ancestors. Thereupon he legitimately succeeds to the status which gives him the ownership of the granary and all that goes with it.

I am concentrating on the first-born son, but two qualifications should be added. Firstly, Tallensi understand quite clearly that he is singled out by reason of his place in the sibling group so to speak; he is, they say, the nearest to the succession. Secondly, first-born daughters have parallel avoidances in relation to their mothers and to some extent to their fathers. This shows how critical is the position of the first-born irrespective of sex. As Tallensi point out it is the first-born whose birth transforms a married couple into parents once and for all.

But I fear that I may be conveying an impression of pervasive tension and antagonism in the relations of parents with their first-born children, the children whose fate it is to have conferred parenthood upon them and to be waiting for the succession. I do want to stress again that there are no obvious signs of anything of this sort in their normal relationships and in their every-day behaviour towards one another. Fathers speak with pride, affection and trust of their eldest sons; rather disconcertingly so when, as so often happens, a father follows a eulogy in the son's presence by adding, 'of course, he is my first-born and though he is still so young he wouldn't care if I died today. He is only waiting to step into my place.' Eldest sons, likewise, as I have already noted, are normally attached and loyal to their fathers. Tallensi are very critical of sons who leave their natal settlement to work or farm abroad for many years. They would be appalled at the idea that a son might resort to violence, or even parricide, as is reported of the Bagisu, in order to assert his claims on his father (La Fontaine 1960).

On the contrary Tallensi never cease to emphasize the duty of

what I have elsewhere (1949, ch. VI) called filial piety; and
that they faithfully observe it is constantly shown. It is illustrated
by the attitude of the youth whose father refused to pay his wife's
bride-price. But similar incidents are of daily occurrence. For
example, I happened to meet Toghalberigu just after a stormy
argument with his father, who accused him of neglecting the family
farm in order to get ahead with the weeding of his own private strip
of land. Complaining to me he ended, more in sorrow than in
anger, 'Is it right, the way he treats me? Yet how can I leave him
since he is almost blind and cannot farm for himself? Would he
not starve to death? Can you just abandon your father? Is it not
he who begot you?'

Here lies the crux. Filial piety is a parent's unquestioned and
inalienable right because he begot you – or, in the mother's case,
she bore you. Character and conduct do not come into it. Bad
parents are just as much entitled to filial piety as good parents. It
is an absolute moral rule. Nor is it purely one-sided; for it is
an equally impregnable moral rule, adhered to with great
fidelity according to my observations, that a parent may not
reject a child, no matter how he misconducts himself. Piety,
in fact, is a reciprocal relationship, compounded of reciprocal
sentiments, ties and duties. And its source (though not its *raison
d'être*) is the irreducible fact of procreation, the fact that confers
parenthood in the elementary sense in which a person achieves
parenthood independently of lineage membership.

What I am calling piety, then, is a complex of conduct and
sentiment exhibited *par excellence* in the relations of a man with
his eldest son and felt to be an absolute norm of morality. It
pervades all their relationships in a curiously interdependent
partnership of growth and development during life. Yet when
Tallensi speak of this relationship they say that the supreme act
of piety required of any man is what falls to him on the death of
his father. It is the duty then of the first-born son, and failing him
of the oldest living son, to be responsible for his father's mortuary
and funeral rites. I am translating the Tale phrase *maal u ba koor*,
for in actuality the elaborate sequence of rites is supervised and
largely carried out by fellow members of the dead man's lineage,
aided by representatives of allied lineages and other kinsfolk.
The children, widows and grandchildren undergo ritual and
observe ritual taboos. They do not officiate.

Nevertheless the essential rites cannot, by right, be carried out without the presence and the lead of the eldest son. And whether or when he takes the necessary steps for this is solely his own responsibility. There are no sanctions of a jural or material kind that can be brought to bear on him. He can, if he wishes, also turn a deaf ear to public opinion, which may be impatient if the deceased held an office, since no successor can be appointed until his final obsequies are performed. Nor may a younger brother take action. That would be usurpation and contrary to the rules of age and generation priority in the sibling group and the lineage. Indeed, it would be an act of impiety against the deceased, even if he had been his father's favourite son. It is wholly a matter of conscience with the responsible son, or as Tallensi put it, it lies between him and his ancestors. If he delays the funeral inordinately, the ancestors will take offence and he will suffer. Funerals are frequently delayed, often for lack of livestock and grain supplies that are required to perform them, but sometimes for motives interpreted by the Tallensi as perverse or selfish. When Nindoghat procrastinated over his father's funeral, among the motives attributed to him were arrogance and malice due to the hostility between his lineage segment and that of the prospective successor to his father's office. So it is not uncommon for diviners to reveal that sickness and deaths are due to wrath of ancestors offended by the delay of a funeral.

Tale mortuary and funeral rites are elaborate and locally varied but here I am concerned only with the most important of those in which the participation of the eldest son is ideally indispensable. (I say 'ideally' because the Tallensi are a practical people and in exceptional circumstances the lineage will act without him.) These are, firstly, the rites by which the deceased is established among his ancestors and is thus transformed from a living person into an ancestor; and secondly and consequentially, those by which the son is invested with his father's status or is made eligible for this. Significantly, it is the eldest son who should make the rounds of all the ancestor shrines that were in his father's custody and, with the customary libation, apprise them of his death. Then, he must attend the divination session at which the ancestral agent of his father's death is determined; for he must concur in the verdict since the sacrifices to appease the ancestors and to reconcile them with the living are his responsibility.

Finally, he (usually accompanied by his first-born sister) is the main actor in rites which free him to do those things which were forbidden to him in his father's lifetime or, in Talis clans, during the period of suspended paternal status since his father's death. No display of grief is permitted in these rites but strict silence is enjoined on the actors in the most solemn of them. This is because the dead is deemed to be participating with them and would strike down anyone who broke the ritual silence.

I want to stop for a moment to consider the implications of these rites. We must remember that among the Tallensi the ancestors constitute the ultimate tribunal, the final authority in matters of life and death. Every normal death is their doing. The deceased is said to have been slain or to have been summoned by them, and it is always in retribution for neglect of ritual service demanded by them or breach of promises made or duty owed to them (*cf.* Fortes 1959). A son is a jural minor during his father's lifetime. As such he has no standing in relation to the ancestors and therefore only indirect and minor ritual liabilities towards them. Thus when he informs the ancestors of his father's death he is, in effect, presenting himself to them as the prospective successor to his father's responsibilities towards them. As heir he must accept the penalty imposed by the ancestors for the fault for which his father incurred death; but though he provides the animal to be offered he may not perform the sacrifice. His father's status has not yet devolved on him. So one of his younger brothers acts as if deputizing for the father.

Then as to the rites of silence, these mime eating and drinking with the dead father, hitherto prohibited to the son. But he must be freed to do so in the future, if he is to be able to sacrifice to his ancestor-father, since this requires partaking of the offering. To end the funeral among the Talis, he takes his father's bow and quiver from the tabooed granary to the external *boghar*. There he hands it to the assembled lineage elders to be deposited among those of the other forebears with whom the father is now joined. Among the Namoos, he is clad in his father's tunic turned inside out, girt with his father's mimic bow and quiver and very solemnly shown the inside of the tabooed granary with gestures that symbolize compulsion from the lineage elders to submit. (*Cf.* Fortes 1949, ch. VIII. It should be noted that the 'bow and quiver' used are small mimic articles made for the funeral and therefore expendable.) At this point he stops being the heir and replaces his father in status. Henceforth he is his own master (within the limits of lineage obligations) jurally, economically, and above all ritually, with authority over his dependants and his family property, and with the right to officiate in sacrifices to the ancestor shrines of which he now becomes the custodian. But he will never cease to be reminded that he holds his status solely in virtue of being his father's successor; for his weal and woe, and that of his dependants hangs upon the will of the ancestors, and they can be influenced only in one way and that is by pious tendance and ritual service, which cannot be rendered except through the intermediation of his dead father.

It is pertinent to add that various offerings of animals and beer are made to the deceased in the course of the funeral rites. They are accompanied by a constant refrain. Always the official calls upon the deceased by name to accept the offering 'in order that you may reach and join your fathers and forefathers and let health, peace, child-bearing, fertility of fields and livestock now prevail'. Thus the climax of filial piety is for the eldest son to see to the proper dispatch of his father to the community of the ancestors of which he now becomes one, and thereupon to displace him in status. It is perhaps not unreasonable or illogical that ancestors, thus dispossessed and thrust out of society by the cruel inevitability of nature, should be known to have a mystical existence, and believed to retain final authority, chiefly by virtue of the pain and misfortune they inflict on their descendants from time to time. No wonder Tallensi declare that it is harder to serve and honour the ancestors with piety than the living. No wonder, too, that they have to find consolation in the belief that the ancestors are always just (cf. Fortes 1959).

Filial succession relates to paternal status in the domestic domain. I have never heard an heir express gratification over this. In fact his attitude, until he is invested with paternal status, is more likely to be one of resignation and submission to what must be. But a lineage successor to office may and does take pride in it. It is rash to speculate about underlying motives in customary behaviour as I hinted at the beginning of this lecture. But I do not think it is going too far to see a connection between these attitudes and, on the one hand, the avoidance relationship between father and son, on the other the equality of lineage brothers.

In a lineage system like that of the Tallensi, paternal status merges into lineage eldership just as and because the relations arising from filiation become relations of common descent in the next and subsequent generations. This provides the framework for the extension of filial piety from the domestic to the lineage level. It is in fact quite directly mediated by the relegation of the father, when he becomes an ancestor, to the communion of all the ancestral forebears symbolically accessible in a shrine dedicated to them collectively.

IV

We are left with the critical question, which I have so far evaded, but which must at length be faced: why piety? or in language which may sound a bit old fashioned today, what is the function of piety in the context of the kinship and religious institutions I have described?

Piety is a word packed with ambiguity, for us, and not altogether free of derogatory associations. If, as the *Oxford English Dictionary* tells us, it commonly stands for 'habitual reverence and obedience to God (or the gods)' and 'faithfulness to duties naturally owed to parents and relatives, etc.', it also carries overtones of hypocrisy and, to quote the same dictionary, of 'fraud and the like practices for the sake of religion'. This is no doubt an understandable ambiguity in a culture which deems it specially meritorious for outward forms of manners and conduct to match inner states of sentiment and belief. Prevarication and hypocrisy are not, however, confined to our civilization. People of good repute among the Tallensi speak of such practices with contempt. At the same time there is little or no questioning of the sincerity of outward conduct. Morality is what is seen in a person's conduct and actions and these are deemed to be expressions of genuine intention, feeling or belief. This is assumed in Tale religious custom and ritual practices, as it seems to have been in the cultures I shall presently refer to, those of ancient Rome, and of traditional China.

To avoid the flavour of unctuous conformity carried by the English word, I venture to use its Latin ancestor, *pietas*, in this lecture, though not without hesitation. Is it not redolent for many of us of long hours of wrestling with the tedious affairs of the Pious Aeneas? 'We are wearied', confesses even the great Virgilian scholar, John Conington (1872, vol. ii, p. 11), writing in more pious times than ours, 'by being constantly reminded of his piety.' Significantly though, he adds that this 'may be partly owing to our misapprehension of the epithet', for Aeneas's piety 'is not merely nominal; it shows itself in his whole feeling and conduct to the gods, his father and his son'. What adds to our understanding of this 'luckless epithet', as another commentator calls it (Irvine 1924, p. 103 n.) is the observation that it is never applied to Aeneas in the Fourth Book while he is Dido's lover but is restored to him when, to quote this commentator, '*pietas* has conquered self' and he leaves her to inspect his fleet.[1] This marks what Warde Fowler (1911 lect. xvii) describes as the 'taming of his individualism' in the interests of the state; for, he continues, '*pietas* is Virgil's word for religion and religion (not knowledge, or reason or pleasure) is the one sanction of Aeneas's conduct.'

The Tallensi do not have a concept for that complex of reverent regard, moral norms, ritual observance and material duty in the relationship between parent and child, more particularly of son to father, both during the lifetime and after the death of the parent, which I am calling pietas. But they would readily understand the Roman ideal alluded to in Conington's apology for Aeneas. A modern authority contrasts pietas as 'concerned with the circles of family and kinship' with 'fides' which 'pertains to the extra-familial', that is the political, side of Roman life. Originally, we are told, it meant 'the conscientious fulfilment of all the duties which the *di parentes* of the kin group demand'. Later it meant both the dutiful discharge of cult obligations to the divine 'members of the kin group' and 'reverence and consideration towards the living, human members' (Pauly 1950, s.v. *Pietas*). An apocryphal story, recorded by various writers of antiquity as an example of superlative piety, and quoted by most commentators, tells of the daughter who kept her aged father alive in prison with milk from her breasts (Roscher 1902, s.v.

---

[1] The reference is to *Aeneid*, IV, lines 393 *seq.*:

> At pius Aeneas, quamquam lenire dolentem
> solando cupit et dictis avertere curas,
> multa gemens magnoque animum labefactus amore
> iussa tamen divum exsequitur classemque revisit.

The Loeb translation reads:

> But good Aeneas, though longing to soothe and assuage her grief and by his words turn aside her sorrow, with many a sigh, his soul shaken by his mighty love, yet fulfils Heaven's bidding and returns to the fleet.

Pietas). This story would strike Tallensi as bizarre but not fantastic. They would applaud the praise implied in the 'luckless epithet' for Aeneas's filial love and would comprehend the definition of pietas as 'dutiful conduct towards the gods, one's parents, relatives, benefactors and country' (Lewis & Short 1880). And they would agree with the significant implication that it belongs to the realm of parenthood not to that of marriage and sexual life.

This need not surprise us. Considering only Fustel's account and overlooking the differences due to the more complex civilization of the Romans, we can see that their patrilineal descent and patriarchal family system, closely bound up with an ancestral cult, has close parallels with those of African peoples like the Tallensi, the Yoruba, the Thonga, and many others who also have segmentary patrilineal lineage systems and patriarchal family structures inextricably tied to ancestor cults. For the nuclear element in all these systems is one and the same phenomenon, the ambivalent interdependence of father and son in the nexus of final authority versus subordination, identification by descent versus division by filiation, transient possession of paternal status versus its inevitable and obligatory supersession by the filial successor.

However, by all accounts[1] the *patria potestas* of the Roman father in the domestic domain was far more absolute than is that of a Tale, or any other patrilineal African father. Age made no difference, as the example of Anchises in the Aeneid shows. Fowler (1911, p. 414) notes that he is the 'typical Roman father' maintaining his authority to the end of his life, 'to whom even the grown up son, himself a father, owes reverence and obedience.'[2] I do not know if Roman sons were compelled to practise avoidances towards their father, but if Cicero's attitude in De Officiis is typical it suggests considerable shyness and formality in their personal relations.[3] It is difficult to believe that there was not a good

---

[1] This is plain both in Maine's analysis of patriarchal power and Fustel's discussion of authority in the family. I base my comments here on these two authorities but modern studies of Roman Law bear them out (e.g. Westrup 1939).

[2] Westrup, vol iii, pt. IV, p. 255 points out that sons were unable to 'found their own *Sacra* or perform the death sacrifices on their own account' while the father was alive.

[3] e.g. *De Officiis*, I, xxxv, 129. 'In our custom grown sons do not bathe with their fathers nor sons-in-law with their fathers-in-law' (Loeb translation).

deal of latent antagonism in the relations of sons to fathers. The
savage punishment traditionally prescribed for parricide and
stories of fathers who sacrificed sons for the public weal[1] lend
colour to this inference and fit in with the rigorous jural sanctions
that supported paternal authority. It is understandable, therefore,
why the right of filial succession was not only very strictly
entrenched, but was in part enforced as an inescapable obligation
to the *familia* in law and to the ancestral deities in ritual. One
can easily imagine the legal constraint and moral compulsion
that was required to bind a son in loyalty to his father, until
the time when, as Maine puts it, the paternal power was extinguished by his death. For not till then was the son jurally
adult and autonomous, which meant *ipso facto* having the capacity
to officiate in religious rites, as Aeneas does in *Aeneid* v, in com-
memoration of his father's death. Nor is it without significance
that this should be his first important religious act.

In the same way, Tale norms of pietas resemble those of
Confucian ethics in China,[2] if allowance is made for the refinement and elaboration added by literary transmission through the
agency of specialists and scholars. Tallensi would accept the
Confucian ideal of pietas as consisting in 'serving one's parents
when alive according to propriety; in burying them when dead
according to propriety and in sacrificing to them according to
propriety', to follow Douglas's translation (1911, p. 119) or
'according to ritual' in Waley's version of the *Analects* (1938,
p. 89). For propriety and ritual are overlapping categories among
the Tallensi, too. We have already seen that Namoos define the
avoidances of first-borns as taboos, whereas Talis say they observe

[1] I am indebted to Mr L. P. Wilkinson, King's College, Cambridge, for the
following note: 'For parricide, the old Roman penalty was to be sewn up in a
sack with a cock, a monkey and a dog . . . and dropped into the Tiber. It is
noteworthy that in 55 B.C. Pompey brought in a law against parricide which
made it punishable in some cases by exile (but 'parricide' could mean killing any
relative).' Mr Wilkinson also reminds me of the story of Brutus who put his sons
to death for plotting to restore the Tarquins whom he had expelled. Virgil recalls
the story in the *Aeneid* VI, lines 815 *seq.* where he speaks of Brutus as 'infelix,
utcumque ferent ea facta minores' – unhappy even though posterity will praise
him for the deed because love of country must prevail.

[2] In the analysis which follows it will be evident that I have derived much
stimulus from Max Weber's famous study of *The Religion of China* (English
translation, 1951, Glencoe, Ill.), where filial piety is discussed *passim* with characteristic perspicacity and learning. But it has seemed to me to be more appropriate
to adduce original observations and sources rather than to cite Weber in detail.

them out of propriety; and this is typical of many Tale religious ideas and practices.

The common ground lies again in the basic similarities between Tale patrilineal descent group and family organization, with its religious projection in ancestor worship, and that of the Chinese. Granted the differences in range and scale due to the greater complexity of a literate, economically and socially differentiated, technologically sophisticated, culturally wealthy and historically orientated civilization, these similarities are noteworthy. Classical treatises on ethics and ceremonial (Douglas 1911; Waley 1938; Grube 1910; De Groot 1910)[1] are echoed in the field observations of sociologists and anthropologists of today. All testify to the Chinese veneration of pietas (*hsiao*) as the supreme virtue in the relationship of children to parents, especially of sons to fathers, both in life and after the elevation of the father to ancestorhood.

Filial piety, in the modern Chinese community described by Hsu (1949, p. 207; and see Freedman 1958) is said to be the 'foundation stone' of its social organization and is given exactly the same sense as in my quotation from the *Analects*. But what is more to the point is the weight given by recent students, also in accord with classical treatises, to the patrifilial nexus in the Chinese family and descent system. 'The basis of kinship is patriliny,' states Hsu (1949, p. 58), 'and the most important relationship is that of father and son. The father has authority of life and death over the son, and the son has to reverence and support his parents. Mourning and worship after the death of the parents are integral parts of the son's responsibility.' Tale fathers do not have unrestricted authority of life and death over their children, but with qualifications this formula would be acceptable to Tallensi.

Furthermore, it appears that in China, too, the first-born son has a unique place in the sequence of the generations, though it is not, apparently, marked by avoidances. Thus it seems that in former times 'the eldest son being the direct propagator of his father's line, had the sole right to make sacrifices to deceased parents. This was associated with pre-eminent rights of inheritance

[1] I venture with diffidence to add that the most cursory reading of such treatises as the *I Li* and the *Li Ki* in Legge's translations (in the *Sacred Books of the East*) is enough to make one realize how the idea of pietas pervaded Chinese thought and life from early times. See the discussion of this topic in Waley.

in regard to property and ancestors' (Sing Ging Su 1922, p. 37). Judging from the references to primogeniture in recent literature, this rule still prevails (*cf.* Freedman 1958, p. 82). We learn from Dr Lin's fascinating story (1948, ch. XII) of the fortunes of a Chinese lineage that the first-born has a 'legal right to an extra portion of the joint property as a special recognition of his primogeniture', and, incidentally, that this may give rise to serious conflicts in the family.

The general impression one forms is that fathers treat their sons with affection and indulgence during infancy, but with increasing authority and formality as they grow to manhood. The emancipation of jural majority, economic independence and the ritual autonomy demonstrated in the right to perform sacrifices to the ancestors comes (as in ancient Rome) at length only after the death of the father, as Hsu specifically states (1949, p. 209). This coincides with the dead father's establishment as an ancestor; and it is noteworthy that tablets dedicated to ancestors are so arranged that those of fathers and sons are on opposite sides of the ancestral hall, successive generations being thus kept apart after death, as they were divided by degree in life, and alternate generations being grouped together in accordance with the well-known principle of the merging of alternate generations[1] (Granet 1951, pp. 86 *seq.*).

I have digressed from Africa to look briefly – and I fear too superficially to satisfy the experts – at the two civilizations which are most renowned for the exalted place accorded to the rule of pietas in their schemes of moral and religious values. These are the paradigmatic cases, often discussed by scholars.[2] They have the advantage that more or less formulated doctrines can be examined to see what is meant by the concept of pietas. But what is most instructive in comparing these paradigmatic institutions with their relatively amorphous Tale – and I believe, more generally, African – counterparts is to consider the reasons for their indubitable efficacy amongst all these peoples, irrespective of how much explicit doctrine there is. It is unnecessary to go farther afield in order to propose some hypotheses. Indeed even if I wished to explore the African data more fully there would be

[1] Granet (1951) makes much of this rule.
[2] e.g. in the admirable article, s.v. 'Piety' in *Hasting's Encyclopaedia of Religion and Ethics*.

little profit in it. For I know of only one modern study of an African religious system in which the observance or neglect of pietas has received particular attention and that is Dr John Middleton's impressive work on Lugbara ancestor worship (Middleton 1960).

Among the Lugbara, as among the Tallensi, the Romans, the Chinese, and all other ancestor-worshipping peoples, a man becomes an ancestor when he dies not because he is dead but because he leaves a son, or more accurately, a legitimate filial successor,[1] and he remains an ancestor only so long as his legitimate lineal successors survive. This goes with the rule that ancestors have mystical power only with respect to their descendants and not, for example, with respect to collateral kin. On the other side, a man has no jural authority in his family and lineage, whatever his standing may be in wealth or influence or prestige, if he has no ancestors and until he acquires the status which permits him to officiate in the cult of his ancestors. For, as Dr Middleton demonstrates at length, authority comes not by delegation from those over whom it is exercised but by transmission and assumed devolution from ancestors. That is why jural authority can be acquired only by succession, in these systems.

But let us consider the paradoxes in these requirements. To become an ancestor a man must have sons; hence the inordinate value attached to male offspring; Tallensi say that a man who dies sonless has wasted his life, and the Chinese, according to Hsu (1949, p. 77), compare him to a tree without roots. We might well ask what deeper motives underly this profound desire for sons, but that would be an unwarranted digression. All we have to note is that sons are desired and needed so that the apotheosis of ancestorhood may be attained; but it can be attained only by so cherishing sons that they eventually supplant one. On the other side, legitimate status in family, lineage, and community can be acquired only by being legitimately fathered; but jural autonomy, which is the source of authority and power even in the domestic domain, can be achieved only upon the death of the father and by assuming his mantle – quite literally so among the Namoos.

[1] This more general formula would include sister's sons in matrilineal systems in which ancestors remain ancestors only so long as their *matrilineal* descendants survive. But I am not considering these systems in the present analysis.

But that is not the end. Purely descriptively considered, as Dr Middleton acutely observes, both as father in his family and as elder in his lineage, the man who holds authority in the name of his ancestors is, by that token, subject to their authority and, in the last resort only to that authority. But it is not the same kind of authority as that of a living person over his dependants. It is imputed to account for things that happen to and amongst his descendants.

So we have the paradox that a man may desire to be allotted responsibility towards the ancestors for ills that befall himself and his family, since this is evidence for all to see that he is directly subject to ancestral authority and by that token jurally autonomous, hence entitled to exercise secular authority in family and lineage affairs as well as to officiate in ritual service of the ancestors.

V

I must resist the temptation to expatiate on this point, and return to the question why pietas? and I shall begin with a proposition that must here be stated dogmatically, though there is ample evidence in the literature I have cited to justify it. In the type of social system we are discussing, at any rate within the domains of kinship and descent groups, jural autonomy and authority are highly prized, indeed the most highly prized capacities a man can aspire to, since he cannot reach full adulthood without them. They are, in a sense, scarce goods, since they are attached to exclusive genealogical positions in a descent group. But jural authority is also indispensable to the organization of society; indeed it is the very heart of the social system. And this must be the reason why authority never dies – must never be allowed to die – though its holders of a given time have to, by the laws of nature. I argue that jural autonomy and authority are attributes of fatherhood. Indeed they issue solely from paternal status, in this type social system. We see this at all levels of social structure, for lineage eldership presupposes paternal status in the holder and is modelled on it. It is not an empty metaphor when Tallensi say that a clan or lineage head is the father of the whole group. And paternal status is not only the kingpin of the social structure in patrilineal systems; it is, among the Tallensi, deeply embedded in each person's life experience by upbringing and through daily cognizance of its existence.

Thus when we say that jural authority never dies and must not be allowed to die we can translate this to mean that fatherhood never dies and must not be allowed to die though fathers in the flesh have to die. There is, of course, a very tangible sense in which fatherhood never dies where patrilineal descent is the governing principle of social structure. For as long as a man's descendants last, their place in society and their social relations to one another and to the rest of society are ordered by reference to his paternal career. They are himself, replicated by social selection as well as by physical continuity. But I am here concerned with the moral and religious representations of this fact. From this position, ancestorhood is fatherhood made immortal, in despite of the death of real fathers; that is to say, it is paternal authority, above all, that is made immortal and impregnable in despite of the transience of its holders. But fatherhood which confers the capacity for authority is worthless without its antithesis of sonship; and sonship is meaningless without the right to attain the coveted status of fatherhood. So we see father and son bound to each other in ineluctable mutual dependence, one might even say in tacit collusion, to maintain this precious value, yet inescapably pitted against each other for its eventual possession, as Tallensi recognize.

In this relationship the sons are at a disadvantage, being under the authority they must support, and what is more, restrained by the premise of kinship amity which outlaws strife between kin (Fortes 1949). Nor must we forget that fathers do love and cherish their sons, and at heart most of all their first-borns, as is apparent from the grief of a father whose son predeceases him. Both among the Chinese and among the Tallensi, they do so openly in their sons' formative years, and still, among the Tallensi, behind the façade of the avoidances, in later years, when their own life situation, and Tale beliefs about human nature, prompt recourse to the defence of thrusting first sons out. Can sons do other than strive to reciprocate their fathers' devotion even while, perhaps only inarticulately coveting their status?

But the problem remains how to reconcile the moral imperative of kinship amity with the rivalry of interests between the generations or, to put it from the angle of the individual's life experience and motivation, how to preserve the trust and affection engendered by the life-long reciprocity of parental solicitude and filial dependence against the pull of the underlying mutual antagonism generated in this very relationship of upbringing. It is, in short, a question of resolving the ambivalence that is built into the relations of successive generations in the unilineally-organized descent systems we are considering. We should bear in mind that there is no means of total escape from the family and lineage structure other than by complete severance of all ties with kin and community and the consequential abandonment of all rights and claims to sources of livelihood, jural status, ritual insurance, and political protection. Traditionally, in a society like that of the Tallensi, one could not live in a community except as either a legitimate member of a lineage and a family, or a kinless and rightless slave attached to a lineage and a family and able to survive only by virtue of being accorded quasi-kinship status. There are good structural reasons, therefore, for institutional devices and cultural values that will serve to regulate the potentiality of schism between successive generations.

Ancestor worship provides the medium through which this end is attained. It represents not only the apotheosis of parental authority but its immortalization by incorporation in the universal and everlasting dominion of the lineage and clan ancestors. How subtly the beliefs and ritual practices of ancestor worship lend themselves to regulating the opposition between successive generations can be appreciated from the manner in which Tallensi rationalize it by recourse to the concept of Destiny (cf. Fortes 1959). This enables them to externalize the latent conflict in symbolic guise, and thus to acknowledge it, without destroying the relationship to which it belongs. But the inequality of power and authority is not eliminated. To accept this more is needed than symbolic cognizance of its character. And here pietas comes into play.

Pietas is rooted in the relations of living parents and children, as I have already emphasized. It enjoins obedience and respect towards parents, submission of personal will and desires to their discipline, economic service to them, and acquiescence in jural minority. The tangible reward for keeping the rules is the gratification of parents and kin, and the diffuse approval of society. There is also a moral reward in that pietas towards the living is *eo ipso* pietas towards the ancestors and is deemed to conduce to their benevolence. We might therefore regard pietas as the temporary renunciation of self-interest in order to maintain indispensable social relationships. But I would rather avoid such conjectures and merely say that conformity to these norms is an avowal of contentment with parental authority and power.

The upshot is that sons who might be tempted to rebel and fathers whose patience is exhausted are both kept in check. But this does not wholly rule out the chances of acrimony or discord between them. Tallensi would say that human nature is like that. There are people who resent authority, or evade customary duties, or flout religious precepts. Sanctions are necessary but must not be expected to work unfailingly. That is how it is even with such emotionally and institutionally compelling rules as the avoidances of first-borns. Tallensi certainly perceive how these observances segregate the spheres of father and son and enable them to put an interpretation on their relationship which reduces friction and suppresses open rivalry especially in the all-important matter of rights over persons and property. But though breach of first-born taboos is unheard of, Tallensi say that faithful adherence to them is a matter of the kind of propriety, and the kind of morality which I have called pietas, not of blind fear. And where pietas is wanting sons may turn against their parents and fathers repudiate their sons, in defiance of both sentiment and religion. I have recorded instances elsewhere (Fortes 1949) and, as we might anticipate, it is usually a mature first-born of an ageing father who rebels and breaks away. But I have not heard of a case in which the ultimate sanction of pietas did not eventually prevail. When the father dies the son must – and in my experience always does – return to perform the funeral and assume his inheritance. Kologo's tragic fate was widely cited as an object lesson. He quarrelled with his father and departed to farm abroad. But when messengers came to tell him that his father had died he hurried home to supervise the funeral. He had barely taken possession of his patrimony when he fell ill and died. The general belief was that this was retribution for failure to make up his quarrel with his father. When he came home for the funeral he made submission to the lineage elders and they had persuaded his father's sister to revoke his father's curse. But this was not enough, as his death proved. The diviner revealed that his father, now among the ancestors, still grieved and angered by his desertion, complained to them of his impiety and so they had slain him.

It is pietas then, which makes living authority acceptable. Transposed into ritual form it becomes the pietas towards the ancestors which is the essence of their worship. This corresponds to the continuity between the living and the ancestors that is embodied in the descent group. But there is tangible foundation for this transposition from mundane custom to religious practices and belief. Among the Tallensi, as among the Chinese but perhaps more conspicuously and familiarly, the ancestors, far from being remote divinities, are part and parcel of the everyday life of their descendants. Their shrines stud the homesteads, their graves are close by, their names are constantly cited in social transactions. It is often impossible to tell, when Tallensi speak of a father or a grandfather, whether they are referring to a living person or to an ancestor. In a large expanded family not a week passes without some sacrifice or libation to the ancestors. And the attitude of the officiant in domestic rites of this kind is but a more reverent version of his relations with a living parent.

I have often taken part with the head of the family in rites that seem so informal as hardly to merit the title of religious worship. On the eve of the sowing season, for example, every family head goes round his homestead and his home farm pouring a libation of millet flour mixed in water upon each of his ancestor shrines in turn in order to inform the ancestors of the tasks that lie ahead. He addresses them with deference and pleads for their protection against accidents and for health, fertility, and well-being for all. But his matter-of-fact manner and conventional words might easily mislead an onlooker to see no religious meaning in his actions. Characteristically, when I found one of my friends supervising the sowing of his home farm, he explained that he was late in starting because, as he put it, 'I had to tell my father first.' As his own father was still alive, I asked if he meant he had to inform him first. 'Yes of course I had to tell him,' he said, 'one can't do anything so important without telling one's father. But I don't mean him. I mean my father who became my Destiny, my ancestor.' Pietas towards the ancestors consists primarily in ritual tendance and services in the form of libations, sacrifices and observances whenever they are demanded. The parallels with pietas towards the living is seen, among the Tallensi, in their description of sacrifice as giving food and drink to the ancestors, though they make it clear that this is not meant in the material sense. In return, they say, ancestors 'back up' (*dol*) their descendants.

However, the aspect of ancestor worship which I wish to dwell upon is its value in resolving the opposition, structural and inter-personal, of successive generations. Granted the premises of belief and value, death palpably removes fathers; but it is not assumed to extinguish fatherhood. On the contrary, it furnishes the conditions for elevating fatherhood above mundane claims and commitments. What is more, it provides the occasion for society to compel sons to accept their triumph as a moral necessity and to make up for it by undergoing the ritual exigencies that metamorphose fathers into ancestors. It is reassuring for a son to know that it is by his pious submission to ritual that his father is established among the ancestors for ever. He sees it as the continuation of submission to the authority that was vested in his father before his death.

And let me interpolate that we must not be deceived into assuming that funeral rites are necessary in order to turn a dead person into an ancestor for what are vulgarly thought of as super-stitious reasons. Similar rites are performed on behalf of living men in order to confer office and status. Ancestorhood is a status in a descent structure as Van Gennep (1909) showed and as such students of African religions as one of our former Presi-dents the late Dr Edwin Smith often emphasized (e.g. Smith 1952). The ritual establishment of ancestorhood defines the realm of events and social relations within which the power and authority of ancestors are believed to be displayed.

To go back to what I have been saying, it should be recollected that death is legitimated as the doing of the ancestors. It is they themselves, the fountain-head of authority and the final sanction of pietas, who remove fathers and open the way for sons to suc-ceed. That they cut down fathers in just retribution for conduct which they are believed to regard as impious is consistent with their status. Is there a more effective way of asserting power and authority than by imputing and punishing disobedience?

It can be seen that in these systems a person never escapes from authority. The jural authority of the living father is metamor-phosed into the mystical authority of the ancestor father, backed by the whole hierarchy of the ancestors and the more formidable for that reason. Thus a father's status is held by grace of the ancestors. For all its rewards, it is not an easy office, for it carries not only material responsibilities for dependants but the more onerous ritual responsibilities to the ancestors. But to succeed him need not be interpreted as supplanting the father, but rather as taking over and continuing the office that was temporarily vested in him. It is submission to duty and this divests it of guilt, the more readily so since the opposition between successive generations is not ended but merely transposed to a new level. And it is in some ways more acute, for misfortunes, disease, and death are the lot of mankind and quite unforeseeable. These are interpreted, in Tale philosophy, as manifestations of dissatisfaction on the part of ancestors. The man who holds paternal status is constantly faced with unforeseen demands from the ancestors. The right to officiate in sacrifice to them gives jural and economic power and authority over living dependants. But it also imposes the burden of responsibility for the proper tendance and service of the ancestors. And one can never be sure that one is fulfilling these obligations satisfactorily, as one can be with one's duties to living parents. (*cf.* Fortes 1959.)

The saving grace is pietas. If one conducts one's life to the best of one's lights, in accordance with the dictates of pietas, one can have faith in the justice of the ancestors. What is more, one can accept what comes from them without remorse and in a spirit of submission to authority that cannot be questioned. Hope remains; for expiation and reconciliation are always open to one. This is no more than admitting that one has failed in pietas, a very human failing, and the institutional means are there for reinstatement. To give the ancestors what they demand in sacrifice, service and observance is to submit to their discipline and so to recover pietas.

I do not think that I have embroidered extravagantly on the ethnographic facts. As far as the Tallensi are concerned, a man's ancestor-father and forefathers are, as I have said, believed to be in his vicinity all the time, ritually accessible to him at the shrines dedicated to them. This is not a superstitious fiction. Their authority, protective no less than disciplinary, is intimately felt to be ever present in normal life, just as living parents are. I was reminded of this on an occasion when the Tongraana was discoursing to his elders about a dispute between two clan heads at which he had been asked to give evidence as an authority on native custom. He explained how he had felt obliged to refute the claims of one of the parties, even though he was a kinsman.

'He was lying', said the Tongraana, 'and lies only get you into trouble. I hate deceit. I will not tell lies, cost what it may. If a man has been properly brought up by his father he will not be a liar. When I was a small boy my father used to beat me and beat me if I deceived him or told lies, that is why I do not speak lies. My father will not permit this.'

The 'father' referred to was long dead but the speaker's manner, gestures and affectionate tone of voice made it sound as if he was there in the room, by his side. I have often had this experience with Tallensi.

To take another instance. Teezien was patiently trying to make me grasp the point of the last fruits rites at the external *boghar* at the end of the dry season.

'We provide for them '(i.e. the ancestors) he said,' and beg crops. We give him (that is, the *boghar* personified as the collectivity of all the ancestors) food so that he may eat and on his part grant us something. If we deny him he will not provide for us, he will not give to us, neither wife nor child. It is he who rules over us so that we may live. Supposing you are cultivating your farm and the crops spoil, will you not say that it is your father who let this happen? If you are breeding livestock and they all die, won't you say your father permitted this? If you give him nothing will he give you anything? He is the master of everything. We brew beer for him and sacrifice fowls so that he may eat to satisfaction and then he will secure guinea corn and millet for us.'

[Ancestors, I protested are dead; how can they eat and do such material things as making crops thrive?] 'It is exactly as with living people,' he answered imperturbably, 'If you have a son and you are bringing him up and he refuses to farm, you upbraid him. You say you fathered him with tribulation and here he is refusing to farm what then are you to eat? If he doesn't farm will he ever get himself a wife, will he achieve children? Now if someone does you a favour wouldn't you go and thank him? And if you do someone a favour and he comes to thank you with however small a token would you not do him a favour again?'

I have reproduced exactly as I recorded them at the time, these reflections on fathers and ancestors of two of the most esteemed, sagacious, and well-informed clan heads whom it was my privilege to converse with in Taleland. It should be added that Tallensi can, at a pinch, call upon and make offerings to ancestors, wherever they happen to be, though ideally the right place for this is at the shrines in the home settlement. You go to a cross roads, and squat facing the direction of your home settlement to make a sacrifice to your ancestors if you are away from home. For your ancestors are always available to you.

We can see that fathers are held in mind as if they had never died. And the image in which they are cast is one that accentuates the authority and discipline which they exercised. They are recalled with pious gratitude for the moral scruples they inculcated and the obedience they exacted, but also with affection for the benevolence they showed to loyal sons. This is more revealing since Tale fathers, in reality, very rarely have recourse to corporal punishment and normally have easy-going and tolerant relationships with their growing sons, as I have previously noted.

In this way the concept of ancestorhood and the religious institutions in which it is ritually and socially embodied serve as the medium that enables the individual to keep up his relationship with his father, even after his death, as if he were a part of himself. The father who controlled his conduct during life turns into an internal censor of his conduct when he becomes an ancestor. And this is effective because he is, at the same time, externally available both for the imputation of absolute power and authority, and for acts of appeasement. This provides grounds that seem rational and objective for the rituals of solicitude by means of which a man may hope to control, or at least to influence and certainly to negotiate the changing fortunes of life. Pietas is the bridge between the internal presence and the external sanctity of paternal authority and power.

VI

To forestall inevitable and justifiable criticism, may I say that I have purposely taken a narrow view of my theme. I have tried to concentrate on the hard core of ancestor worship in one type of social system, and more particularly in a single specimen of the type. I have ignored the ramifications and substitutions by means of which the elementary principles of ancestor worship are extended throughout the entire religious system of a people like the Tallensi. I have, for instance, paid no attention to the extremely important role of maternal ancestors in patrilineal ancestor cults. I have also restricted comparative evidence to the minimum needed for my argument. I wish it had been possible to examine matrilineal descent systems with ancestor cults. I can only say that such studies as those of Dr Colson (1954) and Dr Gough (1958) seem to me to confirm my main thesis. Again, Polynesian moral and religious customs and beliefs, especially those of the Tikopia on which we now have such splendid and rich material, throw penetrating light on the nature of paternal authority and its apotheosis. Granted the differences in religious values and beliefs, there is basically a common pattern, in this regard, among Tikopia and Tallensi. The relationships of fathers and sons in Tikopia are characteristically patrilineal and strikingly resemble those of the Tallensi (Firth 1936).

At all events, the course I have followed was chosen in the belief that the best way to arrive at clear hypotheses is to isolate for analysis what is generally agreed to be the nuclear institution of ancestor worship. Ancestor worship is primarily the religious cult of deceased parents; but not only that. For it presupposes the recognition of ancestry and descent for jural and economic and other social purposes. It is rooted in the antithesis between the inescapable bonds of dependence, for sustenance, for protection from danger and death, in status and personal development, of sons upon their fathers, on the one hand, and the inherent opposition of successive generations, on the other. The ambivalence that springs from this antithesis is due to the fact that a son cannot attain jural autonomy until his father dies and he can legitimately succeed him. The ancestor cult permits this ambivalence to be resolved and succession to take place in such a way that authority itself, as a norm and principle of social order, is never overturned. But to accept the coercion of authority throughout life, first as a jural minor and then in ritual and moral submission, without loss of respect, affection, and trust towards the persons and institutions that must be vested with it for the sake of social order might be difficult if not intolerable. This is where the ideal of pietas enters as a regulative and mollifying directive to conduct.

I said that I have purposely eschewed wide comparison. But I cannot refrain from adducing one striking negative instance in support of my argument, though I am well aware that its value is circumstantial rather than conclusive. Dr Stenning's brilliant analysis of the developmental cycle of the family among the Wodaabe Fulani (Stenning 1959) presents a picture of jural and economic relations between successive patrifilial generations which are in radical contrast to those I have described. Fathers relinquish their control over herds and their authority over persons step by step to their sons during the course of the sons' growth and social development. The process begins when a man's first son is born and culminates when his last son marries with the final handing over to the sons of what is left of the herd and what is left of paternal authority. The father then retires physically, economically, and jurally, becomes dependent on his sons, and is, in Dr Stenning's words, to all intents henceforth socially dead. A parallel process takes place with women. With this is associated progressively increasing skill and responsibility in cattle husbandry for boys, parallel growth into marriage after betrothal for both sexes in early childhood, and a pattern of complementary co-operation between the sexes and the age groups that rules out coercive parental authority. Finally, Wodaabe descent groups generally have a genealogical depth of not more than three or four generations.

In this context of social structure, where there is no need for a man to wait for dead men's shoes in order to attain jural autonomy and economic emancipation, the tensions between successive generations that are characteristic of the patrilineal systems I have been concerned with, do not appear to develop. Does this account for the absence of an ancestor cult among the Wodaabe? Or is it due to their adoption of Islam? Dr Stenning does not think it is due to Islam;[1] and I am naturally attracted to this conclusion since it supports the main thesis of this lecture.

A last question. Is pietas bound up exclusively with ancestor worship or does it reflect a general factor in the relations between successive generations that is only mobilized in a special degree and form in ancestor worship? Do the Wodaabe Fulani, for example, have this notion or not? I cannot attempt to answer this

[1] Private communication from Dr Stenning. He does not discuss the question in his book.

question here.[1] But I am reminded of one of the most delicate descriptions of filial piety known to me.

It occurs in that masterpiece of period ethno-fiction, Anthony Trollope's *Barchester Towers*. You may remember the scene. The old Bishop ('who had for many years filled the chair with meek authority') is dying. His son, Archdeacon Grantly, is by his bedside, and the question that is in everybody's mind is troubling him, too. Would his father die before the out-going government fell? If he did the Archdeacon would undoubtedly, to quote the narrator, 'have the reversion of the see'. If not, another would as surely be elected. Clearly and compassionately, as one who is himself no stranger to such an event and to such emotions, Trollope describes the old man's last moments and the son's thoughts and feelings. 'He tried to keep his mind away from the subject, but he could not,' remarks the chronicler. 'The race was so very close and the stakes were so very high.' He lingers by the bedside for a few minutes and then continues:

'But by no means easy were the emotions of him who sat there watching. He knew it must be now or never. He was already over fifty, and there was little chance that his friends who were now leaving office would soon return to it. No probable British prime minister but he who was now in, he who was soon to be put out, would think of making a bishop of Dr Grantly. Thus he thought long and sadly, in deep silence, and then gazed at that still living face, and then at last dared to ask himself whether he really longed for his father's death. The effort was a salutary one, and the question was answered in a moment. That proud, wishful, worldly man sank on his knees by the bedside, and taking the bishop's hand within his own prayed eagerly that his sins might be forgiven him.'

I need quote no more to establish that the Trollopians were deeply imbued with the sentiments and habits of pietas. It can hardly be an accident that so acute an observer as the narrator of this history should depict the Archdeacon's emotions in the setting of a mature son's compunction over his half-hidden ambition to succeed to his father's office. Nor is it by chance that the Archdeacon is shown to pray for his sins to be forgiven rather than for his father's recovery, improbable as that might be.

[1] A study of Japanese religious practices and values would be particularly rewarding from this angle, since they do not appear to have an ancestor cult but have practices that resemble ancestor worship.

Nor is pietas unknown in our present era of unprecedented technological audacity. The other day we learnt from *The Times* (14 April 1961) how the first cosmonaut prepared himself for his spectacular venture. We are told that 'Major Gagarin was brought to Moscow just before the flight and went to Lenin's tomb in the Red Square "to gather new strength for the fulfilment of his unusual task"'. And nearer home, is not the celebration of Founder's Day at institutions like my own college in Cambridge an act of pietas not too remote in spirit from some of the rites and attitudes I have been discussing?

8

Social and Psychological Aspects of Education in Taleland[1]

I. INTRODUCTION

There is no lack of disquisitions on the role of education in the simpler societies. Africa, in particular, has received enormous attention in this connection. Commissions and congresses have delivered judgement on education in one or another African society. Missionaries, anthropologists, itinerant journalists, travellers, Government officials, and innumerable others have vouchsafed opinions on the subject until it has become smothered in platitudes and generalizations. But empirical studies of a sociological or psychological kind in field or school are far from numerous. In this paper I shall attempt, in outline, such a study.

II. THE SOCIOLOGICAL APPROACH

We can commence from two axioms which must be regarded as firmly established both in sociological and in educational theory. It is agreed that education in the widest sense is the process by which the cultural heritage is transmitted from generation to generation, and that schooling is therefore only a part of it. It is agreed, correlatively, that the 'moulding of individuals to the social norm is the function of education such as we find it among these simpler peoples',[2] and, it may be added, among ourselves.

Starting from these axioms, anthropologists have explored the

[1] An outline of this paper was first presented in a lecture on 'Play Activities of Primitive African Children' which I gave in March 1936 at the invitation of Paul and Marjorie Abbatt. Dr Lucy Mair and Dr E. E. Evans-Pritchard have assisted me greatly with their comments and criticism. It was published in its present form as Memorandum XVII for the International Institute of African Languages and Cultures, 1938.

[2] Hoernlé (1931). But this axiom has been formulated by many writers, e.g. Dewey (1916).

conditions and the social framework of education in pre-literate societies. It has been shown that the training of the young is seldom regularized or systematized, but occurs as a 'by-product'[1] of the cultural routine; that the kinsfolk, and particularly the family, are mainly responsible for it; that it is conducted in a practical way in relation to the 'actual situations of daily life'.[2] It has been observed that manners and ethical and moral attitudes are first inculcated within the family circle in association with food and eating and with the control of bodily functions. A good deal of discussion has been devoted, also, to what appear to be overtly educational institutions, such as initiation schools and ceremonies, age grades, or secret societies.[3] It has been proved that direct instruction in tribal history, sexual knowledge, and ritual esoterica is promoted by these institutions.

In this way a good deal of information has been accumulated about *what* is transmitted from one generation to the next in pre-literate societies, about the circumstances of this transmission, and the institutional and structural framework within which it occurs. Of the process of education – *how* one generation is 'moulded' by the superior generation, *how* it assimilates and perpetuates its cultural heritage – much less is known. The problem has, indeed, never been precisely formulated,[4] with the result that alleged discussions of primitive education not infrequently prove to be merely descriptions of social structure slightly disguised.

The problem formulated. Education is a social process, a temporal concatenation of events in which the significant factor is time and the significant phenomenon is change. Between birth and social maturity the individual is transformed from a relatively peripheral into a relatively central link in the social structure; from an economically passive burden into a producer; from a biological unit into a social personality irretrievably cast in the habits,

[1] See Hoernlé, *loc. cit.*; Malinowski (1936).

[2] Firth (1936), pp. 147 ff. This is the most valuable empirical and theoretical contribution to the subject of recent years. There have been various attempts to utilize this principle in planning curricula of schooling in Africa. *Cf.* Mumford (1930) and Helser (1934).

[3] *Cf. inter alia*, E. W. Smith (1934) for a useful commentary on this and other general points referred to in this paper; and Driberg (1932), pp. 232 ff.

[4] I refer to Africa here. Margaret Mead (1928 and 1931) and Firth, *op. cit.* have done much to clarify it for Oceania. Attention is drawn to the problem in E. W. Smith (1934).

dispositions and notions characteristic of his culture. The problem presented by this function of society is of an entirely different order from that presented by the religious or economic or political system of a people. The former is primarily a problem of genetic psychology, the latter of cultural and sociological analysis. In studying education in a particular society we ought, ideally, to be able to take its cultural idiom for granted, whereas the first task of sociological analysis is to discover the cultural idiom. Thus, in Taleland one often finds a pair of small boys disputing with childish earnestness as to who is senior. Unless one has observed the scope of the principle of seniority in the social structure one is liable to dismiss this as mere childish play of no importance.

Education, from this point of view,[1] is an active process of learning and teaching by which individuals gradually acquire the full outfit of culturally defined and adapted behaviour. In this paper I shall try to delineate briefly how it occurs in one African society, that of the Tallensi of the Northern Territories of the Gold Coast. As it will be impossible to compass the whole process of social maturation within the limits of this paper, some of the more conspicuous changes and activities only will be examined.

The sampling problem. My observations were made in the course of a field study of the Tallensi the object of which, in accordance with the usual ethnographical method, was the entire society and its culture. It is impossible in such a case – as other anthropologists have found – to follow up a special psychological problem in a manner commensurate with the criteria of experimental research in England. That will only be achieved when specialists can be induced to take over these investigations.

Such material as I am able to present consequently suffers most from sampling deficiencies. Social behaviour among primitive people appears to be more standardized than among ourselves. Yet variations occur, distributed perhaps in accordance with the normal curve of error as among ourselves but perhaps

[1] An excursus into genetic psychology would be out of place in this paper, but the reader will observe how much it owes, *inter alia*, to Bartlett (1932); Koffka (1928); S. Isaacs (1930; 1951); Spearman (1923); and to the writings of Jean Piaget (1926, etc.). I have borrowed also from the anthropological writings already cited and from those of Malinowski and Radcliffe-Brown, particularly from their writings on kinship.

not. The problem has still to be investigated. Ethnographers are principally interested in 'patterns of behaviour', hence they neglect or slur over variations, and sometimes indeed build whole theories on a single occurrence. For most problems of social anthropology variations are of minor importance as compared with the 'typical', and an all-round knowledge of a culture is a sufficient check of typicality. For problems of developmental psychology variations may be of the utmost importance. For example, the first case of thumb-sucking I observed was that of a girl infant 3–4 years old whose mother had recently died. Could it be assumed that this was a clear-cut instance of the thumb being substituted for the nipple? Some time later I came across another little girl, about the same age, thumb-sucking, but her mother was alive and she was not yet fully weaned. Further observation brought a few more cases of this habit to light, but it is so infrequent among Tale children that a single year's observation does not yield sufficient instances to suggest any correlation.

Generally speaking, therefore, small samples form the basis of the observations recorded in this paper. Where norms of development are implied it will be understood that an appreciable, though indeterminable, variability exists.

III. SOCIAL SPHERE OF THE TALE CHILD

The process of education among the Tallensi, as among a great many other African peoples of analogous culture, is intelligible when it is recognized that the social sphere of adult and child[1] is unitary and undivided. In our own society the child's feeling and thinking and acting takes place largely in relation to a reality – to aims, responsibilities, compulsions, material objects and persons, and so forth – which differs completely from that of the adult, though sometimes overlapping it. This dichotomy is not only expressed in our customs, it comes out also in the psycho-

---

[1] I am aware that the unqualified manner in which this proposition is formulated here invites immediate objection that it cannot possibly hold for the new-born infant, or even for the child who is inarticulate and does not yet walk or crawl. I would suggest in answer that it is impossible empirically to observe the point at which the child's effective orientation to its cultural environment first receives its bias; and also, that recent psychology and anthropology would support me in attaching basic importance to the earliest infantile responses of a child as determinants of later interests.

logical reactions which mark the individual's transition from the child's world to the adult's – e.g. the so-called negative phase or adolescent instability which has been alleged to be universal in our society.[1] It is unknown in Tale society.[2] As between adults and children, in Tale society, the social sphere is differentiated only in terms of relative capacity. All participate in the same culture, the same round of life, but in varying degrees, corresponding to the stage of physical and mental development. Nothing in the universe of adult behaviour is hidden from children or barred to them. They are actively and responsibly part of the social structure, of the economic system, the ritual and ideological system. Psychological effects of fundamental importance for Tale education follow from this. For it means that the child is from the beginning oriented towards the same reality as its parents and has the same physical and social material upon which to direct its cognitive and instinctual endowment. The interests, motives, and purposes of children are identical with those of adults, but at a simpler level of organization. Hence the children need not be coerced to take a share in economic and social activities. They are eager to do so. This does not mean that Tale children are altogether passive in the hands of their parents. Temperamental idiosyncrasies in children strike the observer at a glance. Tantrums, disobedience, destructiveness, and other aggressive outbursts occur sometimes. Among youths and adults misfits and incompetents can be found. But even they live for the same ends and have the same objective values and interests as the majority.

The unity of the social sphere is the more marked in Tale society since they have almost no social stratification within a genealogical or local community. There are no class or rank cleavages, nor highly organized and exclusive economic, political, or ritual associations cutting across the local, genealogical segmentation; and the social division of labour is rudimentary. The pattern of existence is the same for all members of a given community, varying only in texture or tone from individual to individual. The social sphere is differentiated only as between communities which are generally localized clans or lineages. Thus, the blacksmith lineage of Sakpee is unique, as compared with its neighbours,

[1] See Murphy (1931), pp. 429–32, for a critical discussion of this question.
[2] *Cf.* also Margaret Mead (1928).

in respect to those components of its culture connected with its craft. Only sons of that lineage may be taught the craft of the smith; and even those who through lack of inclination or ability do not learn it share the ideology associated with it and submit to ritual and moral constraints in virtue of this. Because of their traditional craft the men of Sakpee do not eat the hedgehog until they have successfully made a piece of ironwork; and when they were driven from the Tong Hills, abandoned their homes and property but carried with them the sacred anvils. Similarly, the social sphere of the Hill Talis contrasts in many respects with that of their Namoo neighbours on account of their different ritual systems.

Yet it is necessary to note that these differences do not detract from the essential homogeneity of the social sphere for all the Tallensi. They are differences in the substance of some cultural definitions, not in the forms and functions of institutions. The totemic taboos of the Hill Talis are exactly paralleled by the prohibitions of first-born children eating the domestic fowl among the Namoos. What their sacred groves and initiation rites mean to the former, their first ancestor, sacred drums, and the chiefship mean to the latter. The psychological substratum of social behaviour is the same in both groups and in their associated clans, with a sole exception. The initiation cult of the Hill Talis does give a bias to their ideology and so to their social sphere which the Namoos lack.[1] In our present inquiry these differences may be neglected since the dynamic relationships of the individual to his community are the same everywhere in Taleland. Whether a child is trained as a blacksmith or as a farmer, and whether he is inducted into the ritual and ideology of the ancestor cult alone, or into that of the sacred groves as well, the principles are the same.

This background will be continuously discerned in our study of Tale education. But a few clinching observations may be cited here to show that the child is not merely a supernumerary element of his society but thinks and feels himself to be a part of it.

1. I was walking with Samane and his two small sons (8–9 years)[2] across a recently sown field of early millet already a few

---

[1] See Fortes (1936a).
[2] All ages are estimated, but the probable error of such an estimate is at least 12 months.

inches high. I chanced to tread on a shoot. Immediately one of the small boys stooped and carefully raised and replanted the blade of millet. 'Why did you do that?' I asked. 'Don't you know that is our food?' he replied reproachfully.

2. Every small boy of 6–7 years and upwards has a passionate desire to own a hen, and many of them are able to realize this ambition. If you ask a small boy why he wants a hen he will reply in almost the same words as an adult uses to explain the importance of 'things' (*bon*). 'If you have a hen it lays eggs, and you take the eggs and breed chicks, then you can sell the chickens and buy a goat, and when the goat breeds you can sell its offspring and buy a sheep, and when the sheep breeds you can sell its offspring and buy a cow, and then you can take the cow and get a wife.'

3. Maanyeya, a little girl of about 9–10 years, said that she had eaten none of the meat of last night's sacrifice. 'Why?' I asked. When they sacrifice to Zukɔk', she explained gravely, 'women don't eat of the meat. If they do, they will never bear any children, they become sterile.' 'What's that to you?' I said. 'Am I not a woman? Who wants to be sterile?' she responded almost indignantly.

4. I was playing with Tarɔmba and Yindubil (8–9 years), sons of the same joint family, and their friends. We were discussing parents. Tarɔmba and Yindubil, speaking together, told me the story of the latter's mother. Three or four years ago she had run away from their father to marry another man. They said, in the very words that an adult member of the family would use, speaking seriously: 'She ran away and took our belly [that is, she was pregnant by their father] and went and bore over there, so our child is there. Then she bore another child there. When our child is big enough we will separate it [*pɔhg* – the technical legal term] and bring it home.' These two were thus identifying themselves completely with the family, mother-child attachments notwithstanding.

Such examples could be multiplied tenfold, but they will suffice at this stage.

The adult attitude about education. Education is not an entirely unwitting process among the Tallensi. They set a high, and indeed overdetermined, value on their culture, and are fully aware of the fact that its continuity depends upon adequate transmission to their descendants. Their social structure, which is built up on the lineage system, and their economic organization put a premium on this, and a significant motive for stressing the continuity of their culture arises from their ancestor cult. Every Taləŋa desires sons so that there may be some one left to pour libations and make sacrifices to him after his death. The ancestor cult and family morality combine to imbue every man and woman with an intense sense of their continuity, both physical and psychological, with their parents on the one hand and their children on the other. A man feels a moral compulsion to pass on his private possessions, his craft, tools, and knowledge to his son, a woman to her daughter. He has the same feeling about carrying on a craft or cult which had been practised by a parent. Ironwork is not merely a profitable craft to the blacksmiths of Sakpee; it is a religious duty to their ancestors.

The most conspicuous affective moment in the religious system of the Tallensi is this sense of moral obligation to parents and children. It is the counterpart of the notion that the sins of the fathers will be visited on the offspring even unto the third and fourth generation. Last year, for instance, Kuwaas had to promise heavy sacrifices to expiate the sin of his grandfather, who was responsible, three generations ago, for the ravaging of his clan settlement by the Hill Talis.

The idea of education is, therefore, not only understood but also frequently formulated in discussions and conversations. A chief once observed to me that children learn who their fathers and ancestors (*banam ni yaanam*) were by listening at sacrifices. 'Our ancestor shrines are our books', he said. Nyaaŋzum, a man of 45–50, put these conceptions neatly and precisely one day when we were discussing some particularly secret ritual matters. His 'grandson',[1] a youth of about 25, was with us. 'Shall we send him away?' I asked. 'No,' he replied, 'if he listens it doesn't matter. He will not gossip to any one. And when some day I am no longer here, is it not he who will take it on? If he listens will he not know, will he not acquire wisdom?' When children are very small, he explained, they know nothing about religious things. 'They learn little by little. When we go to the shrine they accompany us and listen to what we say. Will they not [thus]

---

[1] Kinship terms placed between inverted commas are translations of native terms indicating classificatory kin.

get to know it?' His own small son Badiwona (6–7 years) is extremely devoted to him. One day, affectionately patting the child, he said: 'When we come out in the morning, and if he doesn't see me, he seeks about for me. If he doesn't find me he won't rest. That is why I had him initiated last year. Whatever I do he also sits and listens. Will he not get to know it thus? And when I am gone will he not say that when he was a child he used to go about with his father and used to see how his father did this and that? This is my child and I am teaching him uprightness. If he is about to do anything that is not seemly I tell him, so that when he grows up he will know upright ways.'[1]

Education, it is clear, is regarded as a joint enterprise in which parents are as eager to lead as children to follow. In consequence of this attitude adults are very tolerant of children's ways and especially about their learning. A child is never forced beyond its capacity. This is seen most clearly in relation to the pivotal economic activity, agriculture. However skilful a boy may be with his hoe, he will not be forced to do as much work as an adult. Men often restrain their sons of 12–14 from joining the adults in hoeing on the grounds that over-exertion is harmful. Again, women do not take daughters of 9–10 on firewood expeditions to the bush.

That skill comes with practice is realized by all. When adults are asked about children's mimetic play they reply: 'That is how they learn.' Thus when a boy is 7 or 8 his father buys him a small bow so that he can go and learn marksmanship in play with his comrades. Yet the existence of individual differences in ability, both amongst children and amongst adults, is recognized and cited with reference to the acquisition of skill. Rapid learning or the acquisition of a new skill is explained by *u mar nini pam*, 'he has eyes remarkably', that is, he is very sharp. A friend of mine who was a cap-maker told me how he had learnt his craft, as a youth, from a Dagban by carefully watching him at work. When he was young, he explained, he had 'very good eyes'. This conception of cleverness is intelligible in a society where learning by looking and copying is the commonest manner of achieving dexterity both in crafts and in the everyday manual activities.

At the same time, no one hesitates to punish when it seems to be merited. A child who neglects a task entrusted to him or her

[1] He learns – *u bamhara*, cognate with *ban* to know; I teach him – *mpaanumi*.

may expect to be rebuked or even chastised. Incidents such as the following occur frequently. Strolling through Puhug one morning in June when the early millet was ripening, I stopped to chat with Tampɔyar, a young man of 25 or so. Suddenly there came the sound of furious railing from his father's house-top. 'Your ears don't listen. Can't you look after your cattle properly, you good-for-nothings, you things-with-sunken-eyes. ...' A voice from a neighbouring house-top joined in to the same effect, adding: 'I'll come down and thrash you.' The objects of these fulminations were two small boys, about 9–10 years, who were driving some cattle out to pasture. They were dawdling and their beasts stopped to munch at the millet. The rebuke had the desired effect. Tampɔyar's comment was to point to a scar on his body where for the same crime he had been so severely whipped, as a small boy, that a suppurating wound resulted.

Disciplinary punishment is also administered at times. A small boy is told to go and scare the birds from the fields. He refuses. A hard smack on the haunches sends him scampering.

Nevertheless, I have never observed vindictive punishment or malicious bullying, either of children by adults or of young children by older ones. Indeed, punishment appeared to me to be extremely rare among the Tallensi, as compared with ourselves. It is thought to be necessary sometimes to use rough measures in teaching morals and manners, but not in teaching skills. A very intelligent elder once declared *nyɛn pu mugx ibii la u ku ŋɔya yam*, 'If you don't harass your child, he will not gain sense.' This view is held by many. But, as the context of conversation showed, what he meant was not so much corporal punishment as that constant supervision is necessary in the training of children for life.

The concept of 'yam'. Tallensi often use the concept *yam* when discussing social behaviour. It corresponds to our notion of 'sense' when we refer to a 'sensible man', or 'sound common sense'. As the Tallensi use the term it suggests the quality of 'insight'. Its range of usage is wide. If it is said of some one *u mar yam pam*, 'he has a great deal of sense', the implication is that he is a man of wisdom, or is intelligent, or experienced in affairs, or resourceful. If some one commits a *faux pas*, or shows lack of understanding, or misbehaves morally the comment is *u ka yam*, 'he has no sense'. The concept is used to refer both to qualities of personality and to attributes of behaviour. It is applied also in a genetic sense to describe the social development of the child. A normal child between the ages of about 6 and 9 years is said to 'have sense at length'. It knows how to behave in the social situations which confront it, whereas an infant of 3 or 4 years old does not. Yet a 4-year-old having learned bladder and bowel control is said to 'have sense' as compared with a 2-year-old. Similarly Nindɔyat, (17–18 years) telling me of his sweethearting, waxed scornful about his own boyish friendship with girls two years before. 'I still had no sense then', he explained. The concept, it will be seen, is generally used relatively to a particular situation.

Children's attitude to education. In Tale society every pupil becomes in some situations a teacher. At the one extreme there are specialized ritual performances which some men do not learn till they are grey, and at the other, the toddler leaning over to play with his nursling brother is teaching the latter something about his environment. This holds right through the age-scale. Considering also the unity of the social sphere, one is not surprised to find that children have the same attitudes about education as their elders. Hence children are rarely unwilling to learn. As a rule, too, they are not ashamed of confessing failure, ignorance, or inability to do or make something. Conversely, children laugh good-humouredly at one another's deficiencies in skill or knowledge, they never jeer. Ridicule is reserved for the correction of uncouth manners, disgraceful morals, or aberrant interests.

Fundamental educational relationships. It will be evident from the preceding analysis that Tale children receive their education not only from the adults but also from older children and adolescents who are always transmitting what they know of the cultural heritage to their younger brothers and sisters and cousins. From the age of 5 or 6 years until they become fully absorbed into the economic system children often go about in small groups. These groups[1] usually consist of siblings, half-siblings, and ortho-cousins of close agnatic kinship, from the same joint family or sublineage group. The composition of a children's group depends upon various factors. The most important of these are the following:

Age and mobility. Infants still requiring care are often carried

[1] See Fortes (1936b) for ancillary information about these groups. Also M. and S. L. Fortes (1936) for an outline of their structural context.

about by boys and girls attached to a group of older children. Generally speaking, children of the same degree of mobility tend to go about together. Thus one often finds a group comprising children of about 6 to 10 years. But the stage of social development reached by a child is important in this connection. Hence young children under 10–11, children between 10–11 and the beginning of pubescence, and adolescents all tend to form separate groups.

Sex. Before the age of about 6 years small boys very frequently go about with girls' groups, and small girls, though more rarely, accompany older brothers. After that age as mobility increases and interests diverge the sexes tend to separate. A well-defined sexual dichotomy runs through Tale social life and thought, and the children begin to adopt the cultural definitions of the roles of the sexes in relation to each other about the time that single-sex groups become common. This is associated with a differentiation of their activities on the periphery of the economic system. Thus cowherds are always boys, whereas it is the girls who help with the housekeeping.

The social situation. Yet none of these children's groups resemble gangs. They have no permanent structure. They generally crystallize out in particular situations and the composition of a group depends largely upon the situation. In ritual situations one often finds a somewhat amorphous group of children and young people of a wide age-range. In games and imaginative play, however, it is more usual to meet with small groups of restricted age-range. Pairs and trios of either sex and of nearly the same age, who are almost always siblings or ortho-cousins, often have a more lasting companionship in work and play and provide a nucleus for larger temporary groups, especially in the dry season.

The most fundamental educational relationship of the Tallensi is, however, not that of children to children but of children to parents. It is a complex relationship, as defined in Tale culture, but its principal moments can be readily discerned. On the plane of rational, everyday economic and social activities there is co-operation, friendliness and tolerance, almost equality. This is more marked between father and child than between mother and child. A father is always addressed by his personal name, a mother is always called *mma*, my mother. Affectively, however, a powerful tension exists between parent and child. A parent's authority may not be flouted, though he or she is expected to be affectionate and indulgent. Sin, in Tale ideology, is primarily an offence against the person, status, or rights of the living parents, or the parent-images – the ancestor spirits.

The structure of this relationship makes it possible for children to acquiesce immediately in the commands and teachings of their parents or parent-substitutes. Children are, as a rule, very obedient. If they refuse to carry out an order there is usually some very valid reason – acknowledged as such by both parents and children – such as that it is some one else's turn. But the parent-child relationship is educationally fundamental in another sense – it is the paradigm of all moral relationships.

It is worth noting, by contrast, that the relationship between grandparent and grandchild is the reverse of that between parent and child. In this case equality, compatibility and partisanship, as if in league against the generation between them, are emphasized, more particularly by the joking relationship – mutual ragging among the Tallensi – permitted between grandparent and grandchild.

The social space. How these relationships function in the educational development of a Tale child is determined by its social space.[1] An individual's social space is a product of that segment of the social structure and that segment of the habitat with which he or she is in effective contact. To put it in another way, the social space is the society in its ecological setting seen from the individual's point of view. The individual creates his social space but is himself in turn formed by it. On the one hand, his range of experiences and behaviour are controlled by his social space, and on the other, everything he learns causes it to expand and become more differentiated. In the lifetime of the individual it changes *pari passu* with his psycho-physical and social development.

In Taleland an infant remains confined to the house for the first three to six months or even longer. During this period its social space is extremely restricted. It has effective social relationships with its mother, sometimes its grandmother or a co-wife of its mother, or an older sibling or half-sibling, and its father – in this order of frequency and intensity. At the age of 3 to 4, when

---

[1] This term is employed here for want of an apter one. It has been used in a different sense by Simmel and others.

it is beginning to talk fluently and can run about, its effective range of contacts includes all the members of the joint family and probably those of closely neighbouring related joint families. It is beginning to associate with groups of other related children belonging to its immediate milieu and to know the topography of its parental homestead and its immediate surroundings. By the age of 6 or 7 it is being taken on visits to its mother's brother's house (*ahɔb yir*) and begins to incorporate into the texture of its life its relationships with its mother's people and their settlement. A little later it commences to take a share in very simple economic duties. In this way the child's social space is continuously expanding as it grows older. The educational agencies to which it is subjected become more numerous and more diverse and its experiences more variegated as it participates in an ever-increasing range of social situations. With adolescence a great increase in mobility ensues and a new interest emerges – the opposite sex as potential spouses. This coincides with the beginning of real economic responsibility. When he reaches adulthood, the individual's social space is a function of the entire social structure in its complete ecological setting.[1] He should be capable of appropriate behaviour in any social situation which may confront the normal Talɔŋa. He is a full citizen – which means that he is actually or potentially subject to the whole gamut of constraints inherent in Tale society.

In the evolution of an individual's social space we have a measure of his educational development. In Taleland this is brought about not only by accretion to, but also by differentiation of, his or her field of social activity. Sex, for example, is a differentiating factor of great significance. From childhood on a person's relation to the economic system is chiefly determined by sex. In the kinship system sex operates as a differentiating factor from an equally early age, from the time, in fact, when a child learns to designate like-sex siblings by different terms from opposite-sex siblings. Local, lineage, or family institutions also act as differentiating factors, though not in the same way. This

[1] A person's social space is not equivalent to the entire social structure since it is included in the latter, which is never fully known to or acted on by the individual. Take, for instance, the Tale joint family. As a structural unit it can be easily distinguished and described by the anthropologist; but each member's view of the whole is unique – that of the head of the joint family is different from that of his wife and son – yet all derived from the same single unit of structure.

is a consequence in part of the peculiarities of Tale political structure. Local groupings, clans and lineages are asymmetrically linked *inter se* and with communities outside the Tale area proper by political, genealogical, territorial, or ritual ties. The most striking case of this sort is presented by the Hill Talis who have immemorial ritual links with villages and genealogical groups among Dagomba, Mamprussi, Bulisi (Builsa), Woolisi (Kassena), Mossi, and other neighbouring peoples. Each of the Hill communities has its exclusive associates in these foreign areas. The people of Sii were traditionally associated with the Bulisi, the people of Gorogo with the Woolisi, the Kpata'ar clan with certain villages of Black Dagomba, and so forth. There was in former times, and is even more so now, a constant traffic between the Tong Hills and the outside areas linked with the clans dwelling there – pilgrims coming to the Fertility Shrines in the Hills and Talis paying ceremonial visits to their associates. But the economic and political by-products of this traffic were most important. In this way a range of geographical and social contacts was, and is, available to a member of the Hill clans which no Namoos had, and among the Hill clans themselves the actual contacts varied from clan to clan. The Namoos again have their traditional political associates among the Mamprussi. It is especially interesting to note that the Mamprussi associates of the Hill Talis and the Mamprussi associates of the Namoos belonged to a single political structure; yet the latter would not, a generation ago, have hesitated to overpower and enslave any Hill Talis they encountered defenceless.

Factors of this sort become significant only after adolescence and more so in the life of men than of women; but they are operative even in childhood. The Black Dagomba guests at a Kpata'ar elder's homestead are familiar figures to his children; and one sign of the interest taken in them by the children is the fact that most Kpata'ar youths understand and easily speak the dialect of the Black Dagomba, whereas other Tallensi do not.

The specific genealogical links of a family act in a similar way. In the extreme case, they may alter a person's whole life. The story of Puvɔləmra is typical, though such occurrences are not usual. He was a Kpatia man whose parents had died in his boyhood. For special reasons he had come to live with his mother's brother at Gbizug. There he grew up, and when I met him, a man of 35–40, he was to all intents a Gbizug man, a complete partisan of that lineage, professing its ideology, subject to its ritual restrictions, and practising the craft of leatherwork as he had learnt it from his mother's brother. He will probably return to Kpatia when he succeeds to his patrimony, there to be assimilated to a rather different ritual system, but transmitting to his sons and grandsons the art of leatherwork and the habits of industry learnt from his mother's brother. Many Tale lineages trace peculiar features of their residence or kinship, or the possession of special ritual, medical, or technological knowledge, to an ancestor with a history like that of Puvɔlɔmra.

Finally, the phase of family history coinciding with an early period of a person's education – the status of one or more senior members of his family, or the holding of ritual or political office by any of them – may have a significant differentiating influence on his or her social space. The homestead of a lineage or clan head is the focus of inter-communication both of the constituent parts of that grouping, the sublineages and joint families, and of the lineage or clan with its neighbours, not only in secular affairs but also in ritual matters. This is even more evident with people who hold offices. There is a constant coming and going of people at the house of a chief or clan-head: a complainant has brought a case to lay before the chief; the elders of the clan have come to consult him about the rain; somebody's daughter recently married has deserted her husband and the indignant son-in-law is brought to the clan-head to discuss the affair; a distant cognatic kinsman has brought a goat to sacrifice to the spirit of the supreme ancestor of the lineage; orders have been received that the young men must turn out for rest-house repairs in two or three days; and so forth. Transactions of this sort generally take place in public and the children of the house are always avid listeners. If any one has to be summoned or a message delivered a youth or a small boy is sent.

The members of such a house, especially those who have reached or passed adolescence, have a wider range of direct contacts with the social structure and their culture than their less privileged contemporaries. In Taleland, it is true, these differences tend to get smoothed out in the course of a person's lifetime owing to the uniformity of the culture and the homogeneity and cohesion of the social structure. Every lineage or clan member is at home in the house of the head, and important affairs are never discussed unless every branch of the grouping is represented. Again, eldership and office circulate from sublineage to sublineage, and the classificatory kinship system spreads the range of identification between the units of the structure widely. Nevertheless, the differential influence of this factor is significant from the educational point of view since our emphasis is on the stage of an individual's life-history at which it operates. The effects are strikingly observable both at the time it is operating and especially in the character of the mature man or woman. An amusing instance of the former is the following: Deɔmzeet of Puhug was elected elder under somewhat dramatic circumstances. Next morning, near his homestead I met his small grandsons of about 6–7 years old. Quite spontaneously they started telling me the story of the exciting events of the past two or three days, obviously repeating the talk that they had heard and only half understood in the family circle. They told me what the other elders had said and what the chief had said and how Deɔmzeet had summoned a diviner to consult. 'Deɔmzeet is the father of all of us,' said one of them, 'and he has a big farm.' They ran off to show me the boundaries of the farm he had recently inherited as head of the sublineage. As we passed the family graves they pointed out, solemnly, the tomb of Deɔmzeet's predecessor, who had no doubt been frequently referred to in the last few days. These were all matters which would normally not have occurred within the purview of children of their age.

Instances of the effect on his adult character of such an expansion of an individual's social space during his youth were so numerous as to suggest a general rule. I always found that, allowing for variation in ability and personality, men whose fathers had been elders or office-holders at a time when they themselves were old enough to take an interest in public affairs as spectators or participants were better informed than the average person, and tended to assume the lead in social and ritual activities. The natives themselves recognize this. It often happens that a man who has not had the advantage of such a training succeeds to eldership or office and has to depend upon the advice and assistance of younger men who have been more fortunate in this respect.

IV. THE DYNAMIC CHARACTERS OF TALE EDUCATION

Learning and teaching proceeds within the structural framework and subject to the cultural conceptions outlined in the preceding pages. We must now investigate its dynamic characters. The educational development of a Tale child may be regarded as the gradual acquisition of an ensemble of *interests, observances* and *skills*. What these are and how they are acquired constitutes our next inquiry. It is basically a psychological problem, tantamount, up to a point, to analysing the observations already recorded at a different level of behaviour.

An exhaustive investigation will not be attempted in this paper. The detailed exposition of Tale child psychology which should be the foundation of such an investigation cannot be undertaken here, nor would it be practicable to try to follow out the complex pattern of Tale education in all its ramifications in a limited space. I propose therefore, to discuss only a few of the major trends and significant processes. Learning and teaching is a composite process, a network of interacting factors. For the purpose of analysis it will be necessary to isolate some of the variables, but it will have to be borne in mind that in the actual life-history of a Tale man or woman they occur only in interaction with one another and with numerous others.

The determinants of social behaviour. From the genetic point of view social behaviour is determined by four groups of factors – physiological, psychological, social, and cultural. If we observe a child learning to walk, we can easily distinguish these factors. It cannot start walking until it has reached a certain degree of physiological maturity. As it practises it learns to plan its efforts so as to avoid falls or trying to accomplish too much at a time – a psychological achievement, depending partly on its level of intelligence. Again, a child learns to walk in response, partly, to stimulation, encouragement and even training given by some or all of the people who come into its social space. The most important of these is, universally, its mother or some one who acts as mother. Finally, there will be culturally defined ways of encouraging and stimulating or restraining a child in these efforts.

In Taleland these factors of social behaviour are notably intercorrelated. In general there is a marked parallelism between the trend of physiological and psychological development and that of social and cultural development – unlike our own society in certain strata of which social development lags behind psycho-physical development. A girl of 16 or so in Taleland is not only physiologically mature but has accepted the role of a mature woman in the psychological and social sense. She is married, she has economic duties to perform, she is socially responsible. A boy of the same age is pulling his full weight in the economic system, and it is regarded as entirely reasonable that he should be thinking of marriage and sex life, though it may be four or five years before he finds a wife. It should be added that physiological development among the Tallensi is probably somewhat slower than among ourselves. Accurate age norms could not, as will readily be appreciated, be established, but I have estimated that the average age of walking is about 2 years, of talking about 3 years (though single words like *mma*, my mother, and names of members of the family are used with infantile approximation at about 2 years). As far as one can judge by somatic indications, pubescence in boys commences at about 14 years. Girls, according to my wife's observations, do not as a rule menstruate before the age of about 15–16 years, by which time they have frequently been married for a year or two.

This close correlation between psycho-physical and social development is reflected in the notion held by the natives that physiological growth is a natural process, like the growth of plants and animals. The Tallensi have no elaborate transition rites to mark the passage from one stage of growth to another. Unlike many other West African peoples, they accept the onset of puberty in boys and girls, which they recognize by special terms, with the same casual rationality as they accept the cycle of time itself. Health and well-being of the body must be safeguarded with all the resources of their empirical and magical knowledge, but growth in itself is not a matter of cultural emphasis. It is of the very nature of human life. It emanates from *Naawun*, Heaven, the *ultima ratio* of Tale philosophy, corresponding in this context to our notion of Nature.

In keeping with this conception of physical growth, children are *expected* to acquire, in due course, the elementary bodily skills – sitting, crawling, walking, running, hand-eye-mouth co-ordination in eating, and so forth. There is no deliberate training in these skills. Parents and older siblings take an affection-ate and attentive, though sporadic, interest in an infant's psycho-physical development, but do not resort to special methods of fostering it. An infant beginning to crawl is allowed to practise more or less at will, watched by mother or sister, brother or father, lest it injure itself. An infant beginning to walk is supported for a bit, now and then, by an older child, or a parent, or any one to whose care it happens to be entrusted. There is no such thing as regular training in these skills. The attention given to an infant in these respects is a function not only of its stage of physiological development, but even more so of the practical exigencies of the situation with which the parent or older child has to cope at the moment. It is quite usual to see incidents such as this: a woman is gathering guinea corn-stalks for her fire some 20–30 yards from the homestead. Her infant son of about 2½ years totters along the path to her calling out *mma, mma* (my mother). She appears to take no notice until he reaches her. He stands beside her, clutching her thigh tightly, while she finishes tying the bundle of guinea corn-stalks. She puts the bundle on her head and, as she wants to get back into the house as quickly as possible, swings the infant – by one arm – up to her hip and marches off. If she were not in a hurry she might walk back slowly, allowing the infant to follow her and throwing back encouraging remarks to him.

Since nobody thinks infants need special training by particular persons, the natives have a habit of bandying them about from one member of a family to another, though grandparents, co-wives, and other children of the same mother are the most usual nurse-maids. A Tale child must, from earliest infancy, learn to co-operate to its fullest capacity with the demands of practical necessity: it must learn to adapt itself to such facts of reality as the economic preoccupations of its mother. I have often seen children of about 12–18 months left sitting in the shade for half an hour or longer, quite alone, while their mothers are busy with some household task. A healthy infant will remain sitting thus, almost motionless, playing with a fragment of calabash, a stick, or the sand – for the idea of giving an infant a toy or some attractive trinket to occupy it when it is left alone does not occur to the Tallensi. On the other hand, when an infant clamours for atten-tion there is always some one of its kinsfolk near at hand to render it.

The native point of view about physical growth was well illustrated in the case of two infants belonging to a single extended family which I knew well. The younger infant was a lively, healthy, inquisitive girl of about 2½–3 years old already toddling about and exploring the world. The other infant, a boy about 3 months older, had been puny and ailing from birth and, though obviously intelligent, was chronically fretful and dependent and had not yet begun to walk. In Taleland the number three is symbolical of maleness, four of femaleness. Hence it is said in their folk-lore that boys begin to walk and talk in their third year, girls in their fourth. Yet the glaring discrepancy in development between these two infants caused no anxiety to the parents of the boy, except about his health. They realized clearly the connection between his retardation and his health, but were entirely complacent about the former. That he would walk and talk in due course was taken for granted.

The expectation of normal behaviour. It will be evident from these examples that in the cultural idiom of the Tallensi age is not conceived as a significant factor in psycho-physical or social development. Yet relative seniority determines status and rights even in children's groups and their notion of time is explicit and clear. They think of genetic development primarily in terms of maturity, which is a synthetic concept embracing both physiological and behavioural signs. Growing up, in other words, is the evolution of one's social personality as it approximates closer and closer to the fully grown, mature adult. Just as this point of view precludes deliberate and standardized methods of training children in the rudimentary bodily skills such as walking, talking, and eye-hand co-ordination in eating, so it would be incompatible with a didactic attitude about bowel and bladder control or about sexual habits and knowledge.

This point of view gives rise to a factor of great importance in Tale education, the *expectation of normal behaviour.* In any given social situation everybody takes it for granted that any person participating either already knows, or wants to know, how to behave in a manner appropriate to the situation and in accordance with his level of maturity. An effort to learn is thus evoked as an adaptation to the demands of a real situation. It is not that people expect one another to act with automatic and machine-like precision; for in point of fact Tale culture tolerates considerable elasticity in the patterns of behaviour. We have an analogy in our own culture in the expectation entertained by most adults that little girls want to play with dolls and little boys with toy engines. We act on this expectation whenever we buy presents for children. How exactly this influences the play of Western children has not been investigated, but there can be little doubt that it does. We expect children to know how to play without being taught.

In contrast to this, many Western mothers nowadays do not expect their children to acquire clean habits simply in the normal course of development, and therefore set out to train them deliberately from earliest infancy in these habits. Tale mothers never train their children deliberately to be clean. When they reach the stage of walking and talking easily, children are expected not to defecate indoors in the daytime but to be asked to be taken outside or to run out themselves. A lapse meets with an expression of disapproval and a reprimand: 'You are big enough already. Can't you go outside to defecate?' The child learns in response to the expectation that it is capable of normal behaviour in that respect.[1] Similarly, infants are expected to understand or to be eager to understand the language of adults, and no one would think of using 'baby talk' to them.

As normal behaviour is always expected, no one hesitates to correct a child or adult who behaves inappropriately through ignorance, and the correction is generally accepted with alacrity and ease. If children are allowed to be present at the activities of adults, they are assumed to be interested in and to understand what is being said and done. No one would inhibit his conversation or

[1] The example is intended to illustrate how the expectation of normal behaviour acts but, of course, other factors are also involved in the Tale child's acquisition of clean habits. These habits are learnt gradually, not all at once. Before it can walk or talk, its mother or an older child sometimes takes it outside when it shows signs of wishing to defecate. By the time it is expected to be clean it is being drawn into the play-activities of slightly older children, most of which take place out of doors. The need to adapt itself to the habits and standards of its older playmates is a strong stimulus to the child to learn not only bowel and bladder control but also the rules of etiquette. As the Tallensi have no special sanitary arrangements, but excrete in the open anywhere near the homestead, taking care only to keep a little way from the paths and from any people who may be about, it is easy for a child to learn the adult convention from the example of older children and adults. Clean habits, like other skills, are learnt as organic responses (see below, p. 233). I shall refer to this matter again, in another context, in a later section.

actions because children are present, or withhold information upon which adequate social adjustment depends from a child because it is thought to be too young. Tallensi, therefore, are not surprised at the comprehensive and accurate sexual knowledge of a 6-year-old, though direct instruction in these matters is never given. As with the ordinary skills and interests of daily life, they expect children to want to know such things. *Naawun mpaan ba*, 'Heaven teaches them', they say, or, as we should put it, it is perfectly natural.

It is known that some people learn more quickly and accurately than others, that variations in skill and ability exist. But the idea of precocity or retardation as a quality of a child's character has no place in Tale thought. A child may intrude on a situation where some one of his or her degree of maturity has no business to be and will be categorically dismissed then; but it would never be rated for being 'old-fashioned'. Every child is expected to be eager to know and to do as much as its social space and its stage of psycho-physical development permits. Hence, though it is clearly recognized that knowledge, skill, and capacity for social adjustment grow cumulatively, the Tallensi have no technique of isolating a skill or observance from the total reality and training a child in it according to a syllabus, as, for instance, we train children in dancing, the multiplication table, or the catechism. Tale educational method does not include drill as a fundamental technique. It works through the situation, which is a bit of the social reality shared by adult and child alike.

The conceptions and practices we have been considering constitute the significant factors in education by participation in practical tasks which is often described as the distinctive method of primitive societies and is as conspicuous in Taleland as in every other preliterate country. It may be observed that even in Western society the principle method of education is by participation. A child repeating the multiplication table is participating in the practical activity appropriate to and defined by the school; but measured by the total social reality it is a factitious activity, a training situation constructed for that purpose. The Tallensi do not make systematic use of training situations. They teach through real situations which children are drawn to participate in because it is expected that they are capable and desirous of mastering the necessary skills.

Corresponding to this contrast in method we can observe a contrast in psychological emphasis. The training situation demands atomic modes of response; the real situation requires organic modes of response. In constructing a training situation we envisage a skill or observance as an end-product at the perfection of which we aim and therefore arrange it so as to evoke only motor or perceptual practice. Affective or motivational factors are eliminated or ignored. In the real situation behaviour is compounded of affect, interest and motive, as well as perceptual and motor functions. Learning becomes purposive. Every advance in knowledge or skill is pragmatic, directed to achieve a result there and then, as well as adding to a previous level of adequacy.

The expectation of normal behaviour and organic response operate also in the education of a Western child, e.g. when it is learning speech or manners. In Taleland it is the most effective factor in the inculcation of a wide range of social behaviour from bowel and bladder control to ritual notions, and from economic skills to sexual habits.

Interests, skills, and observances. We have seen, now, that a Tale child acquires the interests, skills and observances which comprise the repertoire of adult social behaviour not in discrete categories but in a synthetic combination. But it is necessary, for analytical purposes, to define such categories of behaviour. They refer to forms of overt behaviour and not to the psychological functions or mechanisms subsumed therein.

Interests. By *interest* I mean simply the observable fact that, according to their level of maturity, Tale children and adults spontaneously show preference for some activities rather than others; that they have a selective orientation in their social space, e.g. reacting to some people more readily than to others; that they obviously seek to satisfy aims and desires in their activities and show a sense of purpose and sustained effort. Food, for example, is one of the dominant interests of Tale children. In 1936 the ground-nut crop surpassed all anticipations in some districts. An elder, referring to this in conversation a week or two before the crop was harvested, said: 'We have a surfeit of ground-nuts this year. Look at the children. They don't care about their mothers, they don't care about their fathers.' When food is scarce, he explained, children are for ever running in to their mothers clamouring for something to eat. Now they are out all day, and if they feel hungry they simply pull up a handful of ground-nuts to chew and are satisfied. Food, indeed, remains one of the dominant interests of a Talǝŋa from childhood to old age, and with the confluence of other interests and motives has a great effect on Tale economic life. Between the ages of 6 and 14 the interest of boys in food becomes linked up with other interests which have been developing during that period, e.g. in learning to use the bow and arrow, in the comradeship of boys of about the same size, in primary economic and technical processes such as herding and hoeing, in exploring the environs of the settlement, and so forth. Thus, a group of boys out herding cattle often have their bows and arrows with them, and when the cattle are grazing quietly they will search around for cane rats or birds to shoot as titbits.

From a very early age – before they can walk or talk – until adulthood children show a marked and explicit interest in the activities of their parents and older siblings and other adults with whom they come in contact. This is due, in part, to the unity of the social sphere. The children thus begin to adopt the cultural values of their society in infancy, as has been indicated above in our discussion of the social sphere. It may, however, be worth while repeating that children express a keen interest in farming, in livestock, in ceremonies, dances and recreations, and in the conspicuous current activities of their households, in their kin-ship relationships, and so forth, from the age of 3 upwards, long before they can actively participate in these affairs. But it is shown quite clearly in their attitudes when they are allowed to be present at adult activities, and in their fantasies as revealed in their talk and play, which revolve round the themes of adult social life.

Another group of interests significant for the educational development of Tale children is connected with their habitat. By the age of 9 or 10 the children are thoroughly familiar with the ecological environment of their clan settlements. They know the economically important trees, grasses, and herbs, e.g., a girl of about 9 once named and showed me nine varieties of herbs used for making soup. They have a fair idea of the gross anatomy of the fowl, small field animals, and larger livestock. But apart from these achievements which lead one to infer the existence of an interest in the natural environment, I was able to obtain some direct evidence thereof. The natives say that small children frequently ask questions about people and things they see around them. However, listening to children's talk for 'why' questions,[1] I was surprised to note how rarely they occurred; and the few instances I recorded refer to objects or persons foreign to the normal routine of Tale life. It would seem that Tale children rarely have to ask 'why' in regard to the people and things of their normal environment because so much of their learning occurs in real situations where explanation is generally coupled with instruction, and because they hear so much adult discussion, in terms of cause and effect, as these are understood by the Tallensi, of the things they are interested in. Yet two examples will prove that even quite small children react with exploratory interest to what is new and unusual. I gave some tiny tin figures of animals to a group of small children to play with. A little boy of about 5 looked intently at the figure of a horse. 'Why does he stoop like that?' he asked, speaking to himself rather than to his companions. But immediately he answered his own question, 'He is eating grass.' On another occasion I observed the 3-year-old son of our cook, pointing to the garbage tin, ask the horseboy: 'Why has that got a lid?'

Again, making clay figures is a favourite diversion of small boys and adolescents. The standard figures are men and women, cows and horses. But a youngster of about 9 or 10 once produced a motor-cycle, explaining that he had recently seen the Agricultural Officer arrive on one. Another boy, a little older, once made a roan antelope, explaining that some two years previously he had seen one of these animals which, fleeing from a hunting party, had entered the settlement and been surrounded and killed there.

I have not attempted to give an exhaustive account of the interests which Tale children develop in the course of their education. I have tried to indicate only some typical interests and their relation to the educational process.

Skills. The acquisition of skills also commences in early infancy, with the child's first experiments in motor co-ordination and speech. Bodily dexterity provides merely the foundation upon which are subsequently built up the socially important skills of adult life. They include not only the predominantly manual skills,

[1] See S. Isaacs (1930).

such as are connected with farming and care of live stock, with hunting, fishing, building, thatching, cooking, housekeeping and gardening, and the technology of specialists, but also non-manual skills, such as a knowledge of the kinship system, of ritual and ceremonial, of economically useful herbs and roots and of others used as drugs to cure illness, of buying and selling, of law and custom – in short, of the whole body of cultural definitions which guide the Taləŋa in his daily existence. Such skills are the end-results of education; and it will be obvious that they represent merely the cognitive and motor aspects of activities rooted in developing interests and fostered by the expectation of normal behaviour. Typical skills will be referred to as we proceed, but one general feature must be noted here. Children in Taleland are remarkably free from over-solicitous supervision. They have the maximum freedom and responsibility commensurate with their skill and maturity. On the one hand they can go where they like and do what they like, on the other they are held fully responsible for tasks entrusted to them. Thus, a 6-year-old is quite often charged with the care of an infant for several hours at a stretch. Girls of 9 or 10 can be seen in large numbers in every market, selling or buying things for their mothers or themselves. Boys of this age and even younger swarm in the markets. One market day I discovered a boy of about 10–11 selling a basketful of leaves used as a food for animals which he had himself collected in the bush near his settlement. He hoped to earn a penny or twopence to buy arrows-heads for himself. Another day a boy of about the same age offered his services to the butcher. He earned a tenth of a penny which he spent on a feed of locust-bean flour for himself and a friend. Similarly, when there is dancing at a homestead, the children of about 6 years and upwards from the whole neighbourhood congregate there and remain till dawn. They might tell their parents where they are going, but would not be sent for to come home.

Numerous instances could be added to illustrate further the freedom allowed to children. Of the responsibility allotted to them in applying their skills and their appreciation of it, one good example may be quoted. I was chatting with a group of boys, aged about 7–11 years, at sunset one day, when Duunbil, the oldest of them, suddenly exclaimed, clasping his hands in a gesture expressing both amusement and consternation: 'Ma! I've forgotten to untie the goats; I'll get a whipping.' 'You'd better run,' one of the others advised, as he dashed away. His companions explained that he had pegged his father's goats out in the morning to graze and that it was his duty to bring them home again. Goats cannot be left out all night lest hyenas catch them. A few days later I met Duunbil again and he told me, with a grin, that he had not managed to bring the goats in till after dark, but that he had been let off with a mild beating.

Children and adolescents share economic tasks in accordance with their skill and maturity, and they have a strong sense of the rights to which they are entitled in consequence. An adolescent has the right to be adequately fed and supported by his or her parents in return for co-operation in agricultural or domestic work. I have known of youths running away from home when they considered themselves to be badly treated in these respects. Children expect to have their services acknowledged in a similar way. Yindɔl, aged about 11, an intelligent and enterprising lad, was already giving his father valuable assistance, both in the care of live stock and in farming. When the great dances came round, Yindɔl wanted a new loin cloth. His father refused to buy him one, whereupon Yindɔl went on strike, neglected the chickens, the donkeys, the goats and refused to go to the bush farm even after a beating. In the end his father had to yield. *Bii la mar buurt* – 'The boy has justice on his side' – he said, when he told me the story.

A similar incident occurred with a man who supplied us with milk for a short while. Money is scarce in Taleland and 2s. 6d. a month is an enviable addition to family resources. Yet one day he came to tell me apologetically that he could no longer sell me milk. Why? His small son, whose job it was to herd the cow, had rebelled. He used always to milk the cow during the day, in the pasture, when he became hungry; but if he was not going to be allowed to do this because the white man must be supplied with milk, he would no longer drive the cow out to graze.

Increasing skill and maturity, therefore, bring increasing responsibilities but also concomitant rewards – that is, ever closer integration into the system of co-operation and reciprocity which is the basis of Tale domestic economy.[1] The unity of the social sphere, the interest of children in the world of adult activities

[1] See M. and S. L. Fortes (1936), *loc. cit.*, for a more detailed description of this.

and the rapidity with which each advance in educational achievement is socially utilized constitute a ring of incentives which help
to explain the eagerness of Tale children to grow up and take their
full place in adult life.

Observances. Finally, there are the observances which every
Taləŋa has to learn in the course of his or her life. These comprise
ritual, moral and ethical values, norms and obligations, as well
as rules of etiquette and standards of correct conduct in general.
Some of these types of observances can be readily identified in
Tale culture. It is not so easy to determine what exactly are
moral and ethical norms and obligations and their place in this
culture. Certainly an exhaustive study of this, the most recondite
and perhaps fundamental aspect of Tale education, could not be
ventured without first precisely establishing the nature of these
observances and how they are maintained. For the limited
purpose of this paper it will be sufficient to select some representative examples of rules of conduct the violation or neglect of
which is repugnant to the cultural idiom of the Tallensi or, otherwise expressed, would be considered blameworthy or disgraceful,
sinful or wicked, uncouth or embarrassing to others. To ask how
they are learnt is to ask how they are experienced by the growing
child at different stages of its development.

To the native these norms have an arbitrary quality as if they
require no validation, other than the fact that normal and equable
social relations would be impossible without them, that they are
indispensable for satisfactory social adjustment.[1] In Taleland, it is
true, learning the correct observances has many points in common
with learning skills, but there is a significant difference. When a
child is learning a skill he has the test of objective achievement to
evaluate his progress and to stimulate him to further effort;
when he is learning how to behave properly the only test of
attainment is the reaction of other people and his own sense of the
adequacy or inadequacy of his conduct. One can observe, in
consequence, that both children and adults are very sensitive to
ridicule of a lapse in manners or morals but are fairly indifferent
to ridicule of poor skill. The contrast between observances and
skills is evident when we compare the way in which a child learns
to model clay figures, gradually perfecting his technique, with
the 'all or none' way, associated sometimes with corporal

[1] The argument of this section owes much to the stimulus of Durkheim (1925).

punishment, in which he is taught to respect other people's property, or the way he is taught his totemic taboos.

Among the Tallensi, as in many other primitive societies, manners and morals are acquired almost as a by-product of the normal social relations between the growing child and the people who constitute his or her social space. The unity of the social sphere and the expectation of normal behaviour have a correspondingly greater influence than in the acquisition of skills and interests. But the critical factor is the tension inherent in the fundamental educational relationships, the authority of parent over child and of senior over junior.

In Tale ideology this is epitomized in the danger attributed to a parent's resentment. If any one who has reached years of discretion behaves in such a way as to incur the resentment and hostility of a parent, some calamity will certainly befall him or her unless he comes formally to beg forgiveness and the injured parent ritually abjures his or her anger.[1] It very often happens that two children of about the same age and members of a single joint family are related as 'parent' and 'child'. They can be seen playing together amicably, squabbling at times, and working together. But when the 'child' is questioned about his playmate, he will say: *nzorumi, ɔn ndɔyam la zugu*, 'I fear (respect) him, because he begot me.' This is not merely a formula. Sinkawol (aged 23–5) used to order his 'father', Kyekambɛ (aged 12–13) about when the lad was helping with the housebuilding and even scolded him if he bungled a task. This was legitimate technical instruction. But he declared emphatically that he dare not strike the boy, *ɔn ndɔyam la zugu*. From the authority of parent over child is derived that of elder sibling over younger. It is equally absolute. A youth or girl has no hesitation in restraining or correcting a younger sibling with a cuff which often sends the latter off howling at the top of his voice. The authority of seniors in general over juniors is accepted by custom (e.g., among herdboys), though not so absolutely and unquestioningly. It is often maintained by force, or in virtue of superior skill. It is intelligible, therefore, why the Tallensi usually say *ti banam yɛl nla*, 'our forefathers' matter is this' when they are asked to account for an observance, and why their

[1] The connection between one particular aspect of Tale morality, their notion of incest, and the background of kinship has been discussed in Fortes (1936b), *loc. cit.*

children say, *bunkɔra la yɛl ti la ŋwala*, 'the grown-ups have told us so'.

Just as all Tallensi, children or adults, recognize the force of authority, even though they sometimes flout it, so they accept the principle of equal rights for equals. One can see this, not only in standardized patterns of behaviour, but, e.g., in the spontaneity with which children share things and activities. When a small boy snares a bird or kills a mouse or lizard he will always share his catch with friends or siblings, apportioning it according to sex and age. A favourite pastime of boys of all ages, from 5 upwards, is wrestling. Sometimes bigger youths stand by to supervise, but always scrupulous fairness is insisted upon, and thus instilled. It is as correct and even praiseworthy to demand what one is entitled to as it is reprehensible to take what one has no right to. A nice illustration, showing incidentally how such principles are sometimes taught, came to my notice during an evening ceremony which had been unduly dragged out. A dozen or more small boys aged from 5 upwards had gathered at the ceremony to wait for a share of the feast which would follow the religious rites. The food was sent out and distributed by an adult, one or two boys of equal size to a dish. Suddenly an elder called out, 'Where is Zuur?' (a small boy of about 5) and rose to look for him. Zuur lay fast asleep in a corner. The old man shook the child roughly to awaken him, and dragged him towards a dish of food upbraiding him, 'Where do you think you will find food after it has all been eaten? You can't control yourself at all.' Zuur, still somnolent, sat whimpering as he ate, while some one else commented that if you want your share of food you must be there to receive it.

The principle of reciprocity which is thus early learnt in association with siblings and age mates is one of the basic moral axioms of Tale social life. If a man refuses to come to the assistance of a neighbour who has invited a collective hoeing party, the latter will retaliate by refusing assistance to the former at a later date. Often at mortuary ceremonies some one, not obliged by custom to do so, will bring an animal to be slaughtered for the dead, 'because when my father died, he brought a sheep to be killed'. A girl of about 10, after her grandmother's funeral, showed me how half her head had been shaved in the customary way. But her spontaneous explanation was: 'Because, at the feast, when they cook the beans they give me a whole dish to myself.' This moral aspect of Tale social relations, as we can now see, explains the rights which increasing economic skill confer upon a boy or a girl. Those who work for you *deserve* their keep.[1]

Like ourselves, the Tallensi recognize that some observances are merely matters of convention, others are matters of conscience, and a great many have elements of both. Their emphasis, however, differs conspicuously from ours. For example, the notion of cruelty to animals as delinquent or reprehensible would never occur to them. Small boys often catch mice or birds and keep them, tied with a strip of strong grass, to play with for days and eventually kill and eat them. It is, indeed, a crude form of nature study, for they learn a great deal about the habits, the anatomy

---

[1] I have not attempted to track the Tale attitudes towards authority and justice to their roots in infant psychology, as would be necessary for an exhaustive analysis. Some of the phenomena which should be observed in this connection are not accessible to study by the behaviouristic methods of the field worker. But it may be worth mentioning that one can readily observe the constraint and force which, at some times, the affection and indulgence which, at others, are expressed in the way parents treat their infant children. From the day of its birth an infant is subjected to the agonizing ordeal of a daily bath in steaming medicated water which is so hot that it becomes rigid with pain on the first douche. One sees infants of a few days to about 12 months old arbitrarily held down on their mother's laps while medicated drinking water is forcibly poured down their throats as they struggle and splutter. A few minutes after an infant has been treated thus roughly it will be lovingly suckled, fondled and caressed by mother or sister, or affectionately dandled by its father. A 3-year-old suddenly frightened runs to its father, or more characteristically to its mother to bury its face between her knees, clutching her thighs, or to snatch at the breast. The same 3-year-old, in a fit of temper, will nag, whimper petulantly, strike its mother with its little fists. No one beats an infant for this. The mother tries to soothe it, or patiently calls some one – husband or co-wife – to come and take charge of the unruly child. This indulgence extends to the excretory processes, and to masturbation, which is overt until puberty in boys. Until it is weaned, at about 3 – when it can walk – it has very complete possession of its mother. After weaning, which is generally mild, the child becomes more detached from its mother, who now resumes regular intercourse with her husband. At this age, or even younger, children are often playfully threatened by their fathers. 'I will slaughter you if you do so and so', but they always seem to react with complete equanimity to this. Yet a boy of 4–5 years, suspiciously eyeing me from the security of his mother's lap, was heard to say that he was afraid I might cut off his penis. The Tale infant, in short, appears to be permitted to gratify all its wish impulses without restraint; yet it must do so in a social world over which it has no control, in which it is a weak dependant. Here, I think, still speaking in terms of superficial psychology, is the germ of the later domination of respect for authority in the child's moral development. It seems probable, from these observations, that the deeper psychology of the process is not unlike that which has been recorded by students of our own children, such as Dr S. Isaacs.

and physiology of the animals and birds they ostensibly maltreat in this way. They are never checked, and the suggestion that this was bad (the Tallensi have no expression, as far as I am aware, for our notion of cruelty) was received with amused laughter. It is a natural interest of children and not a moral problem; and it does not produce a general attitude of cruelty to animals. Both children and adults often show great affection for domestic dogs, for example.

Many rules of conduct are observed 'because it is befitting' (*de narəmi*) – or because non-conformity is 'unbecoming' (*de pu narəmi*) – i.e. it rouses embarrassment or ridicule or public criticism. Thus when greeting a superior, a man must sit with legs flat, crossed at the calves, bare-headed, eyes lowered, whereas a woman must crouch on all fours or sit with thighs close together and legs tucked under the buttocks. This latter posture in women is a matter of modesty. Up to about 9 or 10 years of age a little girl can sit in any way, legs spread out if she wishes, although from babyhood she has learnt to sit with legs tucked under. As she approaches puberty she will often be admonished 'sit properly' if she does not sit decorously. It is unbecoming for women to eat the domestic fowl, and I have often heard little girls of 9 or 10 boast that they are already refraining from it. A woman goes naked until she is advanced in her first pregnancy. After that she must wear a perineal band even at night, in the privacy of her own room, for the sake of modesty. Not to do so, *de mara valəm*, 'is embarrassing' for her and others. Similarly it is indecorous for a young man, especially if he is married, to go about without a loin cloth. His comrades would scoff at him, womenfolk jeer, and his wife be ashamed. Yet the Tallensi have not the slightest sense of shame about the naked body or any physiological functions, all of which they discuss publicly and openly. Ask any one, child or adult, the sex of an infant and the answer will be to open its legs and expose the genitals with the word 'boy' or 'girl'. An infant's excretory processes arouse no embarrassment and meet with no attempts at regulation. As we have seen it is expected to learn cleanliness in due course. The child is made to understand that excreting indoors is a nuisance to those who have to keep the house clean. No one would dream of chastising an infant for this misdemeanour. These are all matters of decency and decorum learnt without formal instructions as direct adjustments to the social space, through the influence of the expectation of normal behaviour.

It is otherwise with the morality of property. A thief incurs disgrace and universal opprobrium, and even small children of 6 and 7 express contempt for one. A child that steals, like one who neglects a task, can expect a severe beating if his father is a conscientious citizen. 'Don't let Tii (a boy of about 6) come and see your things,' said Batiignwol, a little girl of 7 or 8, to me one day, 'he's a thief. He stole his mother's ground-nuts and meat and his father beat him and tied him up.' Theft is immoral as well as criminal. Lying, by contrast, is considered merely foolish and contemptible. It causes annoyance and a liar's comrades distrust him, but one would not punish a child for lying unless it led to serious consequences. Such misdemeanours, nevertheless, go beyond what is merely 'unbecoming'. They are not nice, *a pu maha*.

Finally there is a type of observance which is a matter of conscience rather than of public approval or reprobation. Many of these have a ritual character and most of them are associated with mystical notions. There are food taboos which must be observed from infancy; there are ritual obligations such as those connected with mourning, which a person may escape till he reaches adulthood, but which children of 9 or 10 are fully familiar with and the compulsion of which they experience to the same degree as the adults; and there is the whole body of implicit moral norms which regulate the day-to-day life of the family and emerge in the duties of children to one another and to their parents.

I have already mentioned the prohibition in some clans against the eating of fowl by first-born children. By contrast with the voluntary abstention from fowl meat practised by women, this is a taboo of supreme mystical value for these clans and is exactly parallel to the totemic avoidances of other clans. The Hill Talis, for instance, may not eat tortoise, some may not eat crocodile or python, and others taboo dog. A 5-year-old knows his or her personal or clan taboo, and can state it emphatically. The remedy for a chance infringement of such a taboo is extremely simple – e.g., in the case of the fowl, crushing some fowl droppings in water and giving the child the fluid to drink. Again the supernatural penalty for a breach is of a vague and general kind.

In some cases it is thought that it would lead merely to an eruption of scabies on the head; in others it is said that the offender would slowly lose health and strength. These sanctions, however, are not the effective agencies maintaining the observances. Breaches of these rules are exceptional, both among children and among adults. A Talǝŋa submits to a food taboo in virtue of a configuration of positive habits and dispositions built up in childhood, not through fear of the sanctions. From the time that it is a babe in arms a child is prevented from tasting or even touching any food which it is prohibited from eating. Later it will be called away from where its companions are sitting over a fire roasting titbits of food forbidden to it – 'Come here, you mustn't eat that; it is forbidden.' I once observed a child of about 5 standing aloof while a group of his brothers and sisters were consuming a dish of python meat – a prized delicacy. 'I don't eat python, it is my taboo,' he explained, with an expression of complete aversion.

A vivid example of how food taboos are inculcated was provided by a biographical reminiscence of the chief of Tongo who, as a first-born, may not eat fowl. There was talk in his court-room of crops and famine in the old days, when an elder wondered if a man would eat a forbidden animal if he were starving. The chief, holding with some of his elders that it would be too repugnant even for a starving man to do so, told this story of how he had himself, on two occasions, violated his taboo. When he was a child of perhaps 5 or 6 he was the favourite of his grandfather, as often happens. Some one brought the old man a gift of eggs which he promply roasted. He had to go outside to attend to some matter and he left his small grandson to see that no one purloined the eggs. Now an egg is a potential fowl, and is therefore equally taboo with the latter. The chief described how there was a brief struggle between his conscience and his desire, and how he succumbed and ate the eggs. His grandfather was furious when he returned and the child in trepidation lied, accusing a girl who was in the next room. The girl denied the accusation and the small boy's father was called in to adjudicate. He at once suspected his son, but the grandfather angrily defended the boy, who was now too terrified even to speak. The matter was dropped; but the small boy, afraid of his father, remained with his grandfather till nightfall. Then he slunk out, only to be caught at once by his father, who had been waiting for him. His father said nothing, but grimly called for a feather and a large gourd of water. 'Drink,' he commanded, and compelled the child, now sobbing with terror, to swallow the whole gourdful of water. 'Open your mouth,' said the father, and began to tickle the child's throat with the feather until he vomited. 'And now,' said the father, grasping the child's arms tightly and striking him right and left, on buttocks, back and head, 'I'll teach you to eat fowl and steal and tell lies.' 'That was how I learnt not to eat fowl,' the chief concluded amid general laughter. But, he continued, many years later, at the time of the great famine, he with two companions was on his way to a neighbouring district and stopped for the night *en route*. They were extremely hungry, not having had a proper meal for days, and when they were offered porridge and a cooked fowl ate ravenously. Now two of them tabooed the fowl, but they were too hungry to resist. The meal finished, they rose, when suddenly the other man who was not allowed to eat fowl was overcome with nausea and began to vomit. Immediately he himself also began to vomit. They had not, after all, been able to stomach the fowl.

How strongly quite young children feel the compulsion of a ritual obligation was brought home to me when a boy of about 10–11 was telling me one day about his dead mother. 'I want them to complete her funeral', he said at length, sadly, 'so that I can have my head shaved' (a ritual obligation). 'Why?' I asked. 'Well, is it not my mother?' he said. 'If I don't have my head shaved it is dirt' (ritual uncleanness).

This is typical also of the moral attitudes to the living, to parents and siblings, engendered in the normal course of family life. That is how children learn to be obedient, considerate of others, ready to co-operate, careful of household property, and so forth.

It is evident from the above examples that adjustment to authority and adjustment to equals act together in the child's acquisition of moral observances. But my observations have led me to conclude that authority is effective from an earlier age than the pressure of opinion from equals. From earliest infancy commands, accompanied by acts or gestures, are constantly being addressed to children. 'Go to your father', 'Take this', 'Come here' – as the child is lifted up, or pulled away from some object it must not touch. Until the age of 3–4, when they are walking and talking well, children appear to be indifferent to the presence of other children or indeed of adults other than their parents and members of their own family. I have watched two infants of 2–2½ years old, both walking, remain for over an hour within a few feet of each other without showing the least interest in each other. By the age of 3½–4 the Tale infant has emerged from this stage of egocentricity (as Piaget has termed it).[1] It will now run to join another child of this age or older, and likes to play with younger infants. It has now also become sensitive to public opinion to a degree not far short of an adult, and therefore capable of adjustment to others, as the following incident illustrates: I was chatting with Ɔmara and a few other adult members of his house when his little daughter Sampana, aged about 3½ years, who knew me as a familiar and friendly visitor, ran out to see me. 'Give me some money,' she asked; and her manner, intonation, and posture were so like those of the grown-up women who often half-playfully asked me for money, that the whole company burst spontaneously into loud laughter. Sampana, obviously taken aback, turned tail and fled indoors where we found her sobbing with chagrin. 'She is ashamed at our laughter,' said Ɔmara, still amused. When the child's mother tried to soothe her she first struggled and struck at her, venting her chagrin on her mother. It was fully ten minutes before her mother was able to quieten her by playing with her and distracting her by making her a toy of wet clay. 'They will laugh at him (or her)' or 'His comrades will laugh at him' is the commonest motive alleged for correct behaviour. If you ask a boy of 6 or 7 why he no longer plays with the little girls he says, 'If I do my comrades will laugh at me'; and little girls say the same. Adolescents, talking about their sweethearts, who are usually clan 'sisters', explain why they avoid the girls of their own section – 'because our comrades will find out and laugh at us'; and the adult is equally sensitive to public opinion.

The total pattern and its genetic development. We have seen that these categories of social behaviour are not learnt in isolation one from the other but as patterns in which interests, elements of skill, and observances are combined. The Tale housewife, going several miles into the bush to collect firewood, is using

[1] J. Piaget (1926).

knowledge of the bush tracks and of the best places for firewood acquired over a period of years; but she is impelled also by the sense of duty to her family and her own self-esteem. The Tale farmer's devotion to his agricultural pursuits is due to a passionate interest in land and crops and to a sense of moral responsibility towards his family and his ancestor spirits. His skill is but one of the factors that affect his general efficiency.

These total patterns which constitute the texture of Tale cultural behaviour are not built up bit by bit, by addition, during the course of a child's life. They are present as *schemas*[1] from the beginning. My observations suggest that the course of development is somewhat as follows: at first the child acquires a well-defined interest associated with a postural diagram of the total pattern. The postural diagram is, as it were, a contour map, extremely simplified and crude but comprehending the essential elements and relations of the full pattern. Further experience strengthens and amplifies the interest at the same time as it causes the details of the postural diagram to be filled out, making it more and more adaptable and controllable, producing more discriminatory responses to real situations, and linking up with other patterns of behaviour and with norms of observance. The total pattern is built up brick by brick, like a house, but evolves from the embryonic form.

A simple demonstration of this principle is provided by a child's learning to dance. A favourite amusement of women with infants just beginning to walk is to let them dance. The tiny tot, barely able to maintain an upright posture for ten minutes or so, is set up on its legs. A couple of women – usually the women of a homestead play with an infant thus when they have no work on hand – calling out to it, with laughter and warmth, 'Come on, dance', begin to sing and clap a dance rhythm and execute a few steps. Tale infants respond with every sign of pleasure to such stimulus and by the age of about 3 have learnt, in a sketchy and diagrammatic but specifically recognizable way, the rhythms and the main steps of the festival dances. The 6-year-old has advanced so far that he or she can sometimes join the real dancing of the adolescents. His sense of rhythm is accurate, he learns the

[1] I have taken this concept from Bartlett (1932), pp. 199 ff., though its bald application here hardly does justice to the significance given to it by Professor Bartlett, and to its value for an understanding of primitive education.

songs quickly, and he has the pattern of the dance clearly. But his dancing is extremely crude still. It tends to be mechanical and monotonous, completely lacking the improvisations and variations, the delicate tracery of step and gesture with which the skilled adult fills out the formal pattern. Every year improves the child's style, but even the adolescent has not yet perfected his or her technique. Yet from babyhood to maturity dancing ability grows continuously.

A child's knowledge of the kinship structure evolves in the same way. The schema, rudimentary and unstable as yet, can be detected in the 3–4-year-old. He or she discriminates kinsfolk from non-kinsfolk, equating the former mainly with people living in close proximity. He knows his own father and mother precisely, but already calls his mother's co-wives 'mother'. Similarly, he knows that 'father' is his own father, but that other men – in the first instance those of the same joint family – are also 'fathers', and he knows that the other kinsfolk frequently seen are brothers, sisters, grandfather, grandmother. But he is still incapable of discriminating genealogical differences; he groups people by generation and by spatial proximity. Thus an adult brother may be described as a 'father'. A child learns the fundamental kinship terms and has the idea of distinguishing its relatives according to generation and genealogical distance long before it can couple this knowledge accurately with differential behaviour towards kinsmen. The 6-year-old knows the correct terms and appropriate behaviour defining its relations with the members of its own paternal family and has grasped the principle of classification according to descent. But in practice he still confuses spatial proximity and relative age with kinship, beyond the limits of his own family. The 10–12 year old has mastered the schema, except for some collateral and affinal kinsmen, the terms for whom are known though he cannot describe the relationships.

Biological drives and cultural motives combine to produce variations in the rates of evolution of different schemas. I have not the experimental data to give accurate or even sample norms, but a rough indication is possible. If the 6-year-old is compared with the 12-year-old in respect to e.g. knowledge of the kinship structure, of agricultural processes, of ritual, and of sex life, the 6-year-old is least advanced in knowledge of the kinship structure and ritual, and most advanced in his or her knowledge of sex life, while knowledge of agricultural processes could fall somewhere between these two levels.

V. FUNDAMENTAL LEARNING PROCESSES

In the course of the preceding discussion I have given several indications of how Tale children are taught and learn. But our investigation would be incomplete without a special consideration of the three fundamental processes utilized by the child in its learning – mimesis, identification, and co-operation. These are not the only learning techniques observable among the Tallensi, but they are the most important; and they usually appear not in isolation but in association with one another. They are most intimately interwoven in play, the paramount educational exercise of Tale children.

Mimesis. Writers on primitive education have often attributed an almost mystical significance to 'imitation' as the principal method by which a child learns. The Tallensi themselves declare that children learn by 'looking and doing'; but neither 'imitation' nor the formula used by the Tallensi help us to understand the actually observable process. Tale children do not automatically copy the actions or words of older children or adults with whom they happen to be without rhyme or reason and merely for the sake of 'imitation'. For hours at a stretch mimetic behaviour may be unnecessary, but in certain types of situation it is the child's readiest form of cognitive adaptation.

Mimesis occurs (*a*) as a response to direct stimulation; (*b*) as an adaptation to a situation the child does not know how to deal with on the basis of its attainments at the time; (*c*) in play, when it is rehearsing in fantasy its interests and the life of the world about it. The Tale boy who learns the leatherworkers' or blacksmiths' craft by closely studying his father at his work and listening to his explanations, tentatively repeating the procedure with scraps of leather or iron and gradually perfecting his skill, is no more learning by imitation than our own children do when they learn arithmetic by copying procedures demonstrated on the blackboard. If he does not understand the craftsman's procedure the lad will never learn it.

We have already had examples of mimesis in response to direct stimulation – e.g. when an infant begins to learn to dance. A child's first efforts at talking are constantly stimulated in this way. At the babbling stage its mother or grandmother, or whoever is looking after it, will frequently in playful mockery mimic its babblings, 'What are you muttering there?' says the woman, 'gba-gba-gba ma-ma …', and the infant is thus stimulated to repeat these sounds again and again. I have often observed incidents such as the following: Duun was playing with his little daughter of about two. He called her, 'Kologo-o-o ee!' and she replied 'm!'; and again he called, and she replied; and so on for about five minutes. Then he said, 'Call the dog, wo-ho, wo-ho!' The infant repeated, 'Wo-wo!' Again he called the dog and the infant repeated the call, and so on several times. Some one spoke to him, and turning his attention from the child, he answered. The infant, still influenced by the set of the game, repeated, as well as she could, a word or two, much to his amusement. As we have seen, the expectation of normal behaviour influences parents to talk to infants as if they understand everything.

When there is a pair or a group of children together, the oldest or most self-possessed generally gives the cues to behaviour which the others follow. This is logical, since collective action must be common or co-operative. Tale adults behave in exactly the same way. In an unfamiliar or difficult situation the best adaptation is to copy the actions and words of any one who understands the situation. Small children, whose schemas are still very rudimentary, are peculiarly apt to encounter unfamiliar elements in situations, and therefore readily resort to mimesis. Between the ages of 3 and 6 years, when the child is eagerly exploring its social space, this happens so often that it seems to develop a habit of mimicking older children in whose activities it is trying to participate. Thus, whenever I encountered little groups of children at play, pegging out goats, gleaning ground-nuts, or doing anything else, and asked the youngest, 'What are you doing?', one of the older children would reply, and the youngest repeat this reply in the same words. One evening I met four or five children from a neighbouring compound idling on a path. One was a girl of about 8 who had her brother of about 5 in tow. She noticed another child some way off and called out a message to him. The little brother, equally interested in the other child, called out and repeated the message in the same words and intonation. He was obviously not merely repeating automatically what he had heard, but endeavouring to draw the attention of the distant child and using the same method as his sister. A great deal of language learning goes on in this way.

Mimesis in play will be considered later.

Identification. A striking feature of social development among the Tallensi is the degree to which children overtly identify themselves with older siblings and parents. It is noticeable in children of 5 and 6 and becomes more marked as they grow older. The parent of the same sex is the model according to which the child regulates its conduct and from which it derives its aspirations and values. Though unwitting, the process is unmistakable. Character appears to run in families. An aggressive, loud-spoken man's children tend to become aggressive and pushing; an industrious man's sons apply themselves to work from early childhood; the dishonesty and unreliability of shifty parents tend to be reproduced in their children; and so forth. The social structure of the Tallensi with its emphasis on family and lineage solidarity, and the unity of the social sphere, encourage such identification. If one asks a child, 'Who are you?' the answer is invariably, 'I'm so-and-so's child', or 'of so-and-so's family', accompanied by a manner or posture which leaves no doubt in one's mind as to who is his father or her mother. Nindɔyat of Zubiung, for instance, was notorious for his selfishness, his arrogance, and his insincerity. His son Sapak, aged 6, was his father in miniature – self-assertive, greedy, combative and re-fractory. Lɔyani, clansman and near neighbour of Nindɔyat, was a complete contrast to him and the most popular man in the clan. His children were among the quietest, most respectful, and most good-tempered I knew. Once they were visiting me when Sapak thrust himself upon us. Within ten minutes he had managed to offend one of the other boys and but for my presence they would have come to blows. A similar contrast in character existed between Nɛnaab's small daughters and Saandi's. Indeed, it would be possible to cite a dozen examples of the way parental character reappeared in children, if space permitted.

A child says, 'This is *our* dog, *our* sheep, *our* land, *our* child, *our* wife', identifying himself completely with his family. Some of the examples given in our discussion of the unity of the social sphere show how identification operates to constitute that unity. It is, obviously, the mechanism mainly responsible for the child's acquisition of interests, and therefore generates powerful motives for following out these interests derived from the world of adult activity. Hence Tale children all want to grow up. I often asked boys of 10 or 12 why they were so keen on hoeing though it was far more arduous than herding or playing about with their friends. The question puzzled many, though they were always emphatic about preferring work on the farm to herding or play. But discussion generally resolved their motives into this: 'We want to hoe because we want to be among the men and help to bring in more food'; and girls have the same attitude about domestic work and child-bearing. One day I found a small boy of about 5 struggling with a large goat. 'Where's your father?' I asked. 'Hoeing his bush farm,' he said. 'What about you,' I said, 'can you hoe?' 'Of course I can,' he replied proudly, 'didn't I hoe the whole compound farm?' I teased him a bit, asking him about the details of farming, and eventually he said, 'No, I can't hoe. I'm still too small,' Similarly, a little girl of about 8 told me that she had accompanied some older girls to the bush for firewood the previous day. Her mother burst out laughing and called to me: 'Don't believe her, she's much too small.' But these innocent fabrications illustrate the contents of children's identifications with adults.[1]

Co-operation. Mention has previously been made of children's co-operation in the ordinary social and particularly economic activities in a manner and to an extent commensurate with their maturity at the time. But it must be referred to again in order to emphasize why it is specially important for children's learning. The little girl who goes with her mother to the water-hole and is given a tiny pot of water to carry is making only an infinitesimal contribution to the household's water supply. Yet it is a real contribution. She learns to carry her little pot of water in relation to a real need of the household. The boy at a sacrifice called to hold the leg of a carcass while it is skinned not only receives his first lessons in anatomy thus; he is performing a task necessary for the completion of the ceremony. The children summoned to carry balls of swish for the men building a house are contributing valuable labour. The child's training in duty and skill is always

[1] This exposition of identification is over-simplified in order not to burden the general argument. In Taleland, as elsewhere, it is not the sole determinant of character formation. It has been stressed here in order to underline its educational importance.

socially productive and therefore psychologically worth while to him; it can never become artificial or boring.

Yet as long as the child's co-operation is limited to subsidiary assistance – that is, until he reaches adolescence – it includes an element emanating from the child's own psychology, an element of play. Upon the adult Talɔŋa the economic system bears down with a disciplinary and constraining effect. The house must be built, plastered, and thatched in time, before the rains come; the crops must be planted, hoed, harvested at the right moment, and the penalty for negligence is severe. The child cannot yet experience this. Its co-operation is still partly wish-fulfilment and has for it the attractions of an imaginative experiment. Thus, whenever there is building in progress, they play at building as well as help carrying swish; and the play part is for the child of equal importance with the work.

Play in relation to social development. The concept of play (*ba deemrɔme* 'they are playing') is well defined and clearly recognized by the Tallensi. The play of Tale children, it has been pointed out, emerges partly as a side-issue of their practical activities. It is also an end in itself, and has a noteworthy role in their social development. In his play the child rehearses his interests, skills, and obligations, and makes experiments in social living without having to pay the penalty for mistakes. Hence there is always a phase of play in the evolution of any schema preceding its full emergence into practical life. Play, therefore, is often mimetic in content, and expresses the child's identifications. But the Tale child's play mimesis is never simple and mechanical reproduction; it is always imaginative construction based on the themes of adult life and of the life of slightly older children. He or she adapts natural objects and other materials, often with great ingenuity, which never occur in the adult activities copied, and rearranges adult functions to fit the specific logical and affective configuration of play.[1]

[1] In this respect Tale children resemble our own children; *cf.* S. Isaacs (1930), pp. 99–102. She sums up: 'Imaginative play builds a bridge by which the child can pass from the symbolic values of things to active inquiry into their real construction and real way of working.... In his imaginative play the child recreates selectively those elements in past situations which can embody his emotional and intellectual need of the present, and he adapts the details moment by moment, to the present situation... *Cf.* also the same author (1951), p. 425, 'Play ... is supremely the activity which brings him [the child] psychic equilibrium in the early years.'

A typical play situation. How vividly these motifs appear in the play of Tale children will be evident to the reader if I describe shortly an actual play situation as I observed and recorded it. I shall describe a typical play episode among children at the transition from infantile egocentricity and dependence to social play and participation in peripheral economic activities. It will be noted how recreational and imaginative play are interwoven with practical activities and how infantile habits still persist. The children's interests fluctuate from moment to moment, egocentric attitudes alternate with co-operative play, and the economic task receives only sporadic attention. Later on we shall consider these factors in relation to the phases of development of children's play.

On a morning in June I found Gɔmna, aged about 7, his half-sister Zɔŋ, aged about 6, and his friend Zoo of about the same age out scaring birds on the home farm. They sat astride the trailing branch of a baobab tree on the boundary of Gɔmna's father's farm and Zoo's father's farm. They slid down to talk to me, and a bigger lad, Tɔŋ, aged about 10, joined us. Gɔmna had wandered off a few yards and now came running up with three locusts. He called to his sister and Zoo. Eagerly they squatted round the locusts. 'These are our cows,' said Gɔmna, 'let's build a yard for them.' Zoo and the little girl foraged around and produced a few pieces of decayed bark. The children, Gɔmna dominant, giving orders and keeping up a running commentary, set about building a 'cattle-yard' of the pieces of bark. Tɔŋ, the older boy, also squatted down to help. He and Gɔmna constructed an irregular rectangle with one side open of the pieces of bark. Zoo fetched some more pieces of bark which Tɔŋ used to roof the yard. The little girl stood looking on. Gɔmna carefully pushed the locusts in, one by one, and declared, 'We must make a gateway.' Rummaging about, the boys found two pebbles which they set up as gate-posts, with much argument as to how they ought to stand. Suddenly the whole structure collapsed and Tɔŋ started putting it up again. The little girl meanwhile had found a pair of stones and a potsherd and was on her knees, 'grinding grain'. Suddenly the two small boys dashed off into the growing grain, shouting to scare the birds. In a minute or two they returned to squat by the collapsed 'cattle-yard'. They appeared to have forgotten all about the 'cows', for they were engrossed in a conversation about 'wrestling'. Some one called Tɔŋ, who departed. Zɔŋ, finishing her 'grinding', came up to the boys with the 'flour' on a potsherd and said, 'Let's sacrifice to our shrine.' Gɔmna said indifferently, 'Let Zoo do it.' Zoo declared that Gɔmna was senior to him, and an argument ensued as to who was senior. Eventually Gɔmna asserted, 'I'm the senior.' Zɔŋ meanwhile had put down her 'flour' which was quite forgotten; for Zoo challenged Gɔmna's assertion. Gɔmna retorted that he was undoubtedly senior since he could throw Zoo in wrestling. Zoo denied this, and in a few minutes they were grappling each other. Gɔmna managed to throw Zoo and rolled over him; but they stood up in perfectly good temper, panting and proud. Suddenly with a shout Gɔmna began to scramble up the baobab branch, followed by Zoo, calling out, 'Let's swing.' For a minute or two they rocked back and forth on the branch and then descended. Now Gɔmna remembered his cows. Vehemently he accused his sister of having taken them, and when she denied this challenged her to 'swear'. 'All right,' she said placidly. Gɔmna took a pinch of sand in his left hand and put his right thumb on it. Zɔŋ licked her thumb and pressed down with it on Gɔmna's thumb-nail. He stood still a moment, then suddenly withdrew his thumb. (This is a children's play ordeal.) Gɔmna examined his sister's thumb and found sand adhering. 'There you are,' he said, rapping her on the head with a crooked finger. The 'cows' were completely forgotten though, for now they turned their attention to me, asking me various questions. After a while Gɔmna, who had been observing my shoes, said, 'Let's make shoes,' and took a couple of pieces of the decayed bark previously used to build the cattle-yard to make shoes. He and Zɔŋ found some grass and old string lying about and tried to tie the pieces of bark to the soles of their feet. Gɔmna now noticed his 'cows' and picked them up, but he was still trying to make 'shoes'. The 'shoes' refused to hold together so he abandoned them and squatted over his 'cows' for a moment, moving them hither and thither. 'I'm going to let them copulate,' he burst out with a grin, and tried to put one locust on top of another. Looking up, he noticed a flutter of birds' wings. 'Zɔŋ,' he cried, jumping to his feet, 'scare the birds!' and he raced into the grain, followed by his companions, shouting and stooping to pick up handfuls of gravel to fling at the birds. For the next five minutes they were engrossed in bird-scaring. The entire episode lasted over half an hour.

The developmental phases of play. Infant play. Up to the age of 6 or 7 a good deal of play, especially that of boys, consists in sheer motor exuberance. Small boys run about, leap and prance for the pleasure of it, frequently in a totally unorganized way. But even 3-year-old boys often introduce mimetic themes, spontaneously or in response to suggestions from older children. They 'ride horses', using a long stick as a horse, with a wisp of grass attached as bridle. In the Festival season they love 'playing drums'. Cylindrical tins discarded by our cook and useless as receptacles were in great demand for this purpose. A remnant of goatskin tied over one end with a strip of bark or grass makes a satisfactory diaphragm, and a hooked twig serves as a drumstick. Often an older boy of 8 or 9 manufactures a toy drum for himself or an infant brother. A small discarded calabash is covered with a piece of skin – a remnant of goatskin or the skin of a rat caught by the boy himself and prepared by himself. The skin is cleverly attached to the calabash with strong thorns. Small boys delight in walking round, tapping out a rhythm on a toy drum, sometimes executing a few dance steps. Another instrument they like to copy is the *kolog*, a single-string fiddle. I once saw a boy of about 6 singing to a 'fiddle' which consisted of a large bow made of a green stick and a thread of grass – the 'fiddle' – and a smaller bow of the same materials – the 'bow'. A bow of the very same sort with a short piece of thick reed as arrow is the 4- to 6-year-old's introduction to the handling of bow and arrow. They very soon develop an accurate aim at a dozen yards or so. Play of this kind is generally very egocentric. I have watched groups of children playing side by side – a boy with a 'drum' absorbed in his banging, another lying on his back absorbed in fantasy, a couple of little girls playing at housekeeping – all indifferent to one another's activities.

The little girl of 3 to 6 plays in much the same way at times; but she is already being drawn into the family play of slightly older children, and, like the small boy with the toy bow, she tends to mimic simpler features of older girls' play when she is playing alone. Hence one often sees a little girl of this age sitting and playing at 'grinding millet' – one stone as metate, a smaller stone as muller, and a handful of sand or a potsherd as the grain.

Play in early childhood. Between the ages of about 6 and 10 the play of both sexes becomes more social and more complex. This is the period when the child is beginning to co-operate in real economic activities, subject to real responsibility, and is acquiring a knowledge of his social space. His or her play reflects these experiences and reflects also the interests and activities character-istic of the stage of maturity just ahead of it. There is, in all Tale children's play, this feature of looking ahead, as it were, experi-menting tentatively with what lies just beyond the present psychological horizon.

During this period the younger children, both boys and girls, have charge of the goats, scare the birds from the newly planted fields and the crops, run errands, nurse the infants, and so forth. The boys help in sacrifices to domestic shrines; the girls assist in household tasks such as sweeping and carrying water. Towards the end of this period boys whose fathers own cattle go out with the cattle-herds and girls are beginning to help in the preparation of food. A 10-year-old girl can prepare the majority of usual dishes. At the age of 6 or 7 brothers and sisters often play together; at 9 or 10 the sexual dichotomy has become firmly established.[1]

At this stage the play of infancy develops in three directions. The sporadic motor exuberance is transformed into recreational play – organized group games and dances; the rudimentary mimetic play becomes elaborate and protracted imaginative and con-structive play; and the rude toy-making of infancy grows into children's arts and crafts.

Tale children have a great many organized games, passed on from one generation of children to the next by drawing younger ones into the games played by those who already know them. The games are traditional, and often built round themes derived from the cultural idiom – farming, hunting, marriage, chiefs, etc. But their value is predominantly recreational. Children play them for the pleasure of collective singing, rhythmical physical activity, and sensory and bodily stimulation. The ordinary dances of moonlight nights in which both adults and children participate are regarded as 'play' of this kind.

Kuobon is a game of this sort. Both boys and girls from 7–8 to about 15 years play it on moonlight nights in single-sex groups. A group of about the same size forms a ring, clasping each other

---

[1] Nevertheless, this sexual dichotomy is not so absolute as to prohibit a girl or a woman from stepping into a breach if there is no male available for a man's task; and men often undertake women's work in an emergency.

and standing on one leg. The other leg is extended towards the centre of the ring, and the children arrange themselves so that their extended legs cross one another. The group then commences to revolve, singing as they go round and round, *kuobon, yee-e-e la ŋaah*, 'fruits of farming, yee, how nice'. The game goes on until they are tired and the extended legs begin to drop. It breaks up with much laughter, only to start again after a rest. Similar games are played by boys only and others by girls only; but for both sexes dancing is the supreme recreation. In both the mixed dances and the single-sex dances one invariably sees a few small boys or girls at the tail end of the line. At the beginning of the Festival season one often meets a group of children about 6–9 years old practising the dances in their play, crudely but recognizably.

At this stage, and till pubescence, boys spend a great deal of their leisure improving their dexterity with the bow and arrow. They now have bows like a man's but smaller, and real arrows with unbarbed heads. They go about in small groups practising marksmanship – shooting at a guinea corn-stalk or a chunk of wood. Sometimes they challenge one another and shoot according to certain rules. The loser forfeits an arrow to the winner. All this is recreational play; but it has a very practical aspect, recognized by adults and children. In Taleland the bow is the symbol of manhood; and every man must know how to wield it. The long years of practice necessary to become an accomplished shot begin with the small boy's first toy bow and extend through the play of childhood and pubescence. Part of this play-practice is the hunting of small field animals and birds. To the boy it is a real hunt, demanding knowledge and alertness and yielding a favourite titbit. Yet it is play as well, being neither obligatory nor dangerous, and being mimetically derived from an adult activity. This has great educational importance. The boy in his play identifies himself with the men, accepts their practical valuation of the bow and arrow, and tries out, as it were, what it feels like to be a man in this respect. Boys often hunt thus in groups, especially when they are out herding cattle, and share the spoil, thus training themselves in co-operation and fair dealing. By the age of 11–12 boys begin to accompany their fathers or elder brothers to real hunts, though they remain onlookers for the most part, whose principal task is to help carrying home anything killed. Not till adolescence will they be allowed to use barbed and poisoned arrows; but quasi-playful hunting thus shades over into the real activity for which it is a preparation.

Imaginative play is rich and frequent during this period, though its themes appear to be few. Family life, the principal economic activities, and domestic ritual supply the mimetic content. Sometimes children are entirely preoccupied with such play for hours at a time; often it is interwoven with practical activities or appears as a resonance of practical activities in which the children co-operate.

On any day in the dry season or the first half of the rainy season one can find a group of girls playing at housekeeping. Most commonly they consist of a group of sisters and ortho-cousins – two to four active participants with, perhaps, a couple of infants attached. Often one or two small boys of about 5 or 6 are in their company, sharing in their play or absorbed in their own separate play. In play, as in the simple economic duties, there is as yet no marked sex dichotomy at this age; and small boys are not ridiculed for 'grinding grain' and 'cooking porridge', or small girls for 'building houses' in play. The girls generally constitute a mixed age group, varying from 6 to 10–11 years, for even those who are already capable of real cooking enjoy playing at it. When they are of about equal maturity their play tends to be loosely organized. Each cooks for herself, but they help one another, lending one another 'utensils', 'grain', 'firewood', and exchanging 'dishes of porridge' like co-wives. When one girl is older than the others she tends to take the lead, and the smaller girls assist her on the pattern of daughter helping mother in real cooking. Infants are 'our children'; and reliable informants have told me that small boys are said to be the 'husbands' – but I have never observed boys being addressed thus, though I have watched housekeeping play very frequently. According to my observation, it is merely implied in the manner of distributing the 'cooked porridge', which follows the pattern of family feeding. Older girls sometimes introduce dolls as the 'children' – clay figures of people made by themselves or, more usually, for them by their brothers.

The essentials of the play consist in 'grinding flour', 'cooking sauce' and 'porridge', and 'sharing out' the 'food'. Every feature of the real processes is mimicked, but with the most ingenious imaginative adaptations. A pair of flattish stones or a boulder

and a large pebble serve as 'grindstones'. For pots, dishes, calabashes, and ladles various things are used – old sherds chipped
into roughly circular pieces the size of a half-crown or crown,
fragments of old calabashes, the husks, whole or bisected, of
mɔlɔmɔk or kalɔmpoo, spherical fruits of common trees varying
in size from that of a large marble to that of a cricket ball, and
even old tins or bits of tins, while some girls make little pots of
clay. Pebbles make a fire-place, a thin piece of millet-stalk is the
stirring-stick, some dried grass the firewood. Sometimes a real
fire is lit, but usually it is merely imagined. Real grain is never
used in such play – it is too valuable to waste thus, as the children
themselves would be the first to insist. A piece of potsherd
pounded up or a handful of sand serves as grain; but a much more
realistic effect is sometimes achieved by using dry baobab stamens.
These can be 'winnowed' and the 'grain' ground. Green weeds
and leaves are vegetables.

The children play with great zest and earnestness, yet never
forget that it is but play. As they grind they hum in a low voice a
grinding song they have heard from mother or elder sister. They
examine the 'flour' to see that it is fine enough, try to get the
right proportion of water, stir the 'porridge' thoroughly, 'dish'
it out with scrupulous fairness. There is a constant interchange of
conversation and commands to the smaller children: 'Bring me
that dish', 'Lend me your broom', 'That's my firewood', 'Don't
stir so fast', 'Come and fetch your porridge', and so on. As a rule
they play together most amicably. I have observed arguments in
such groups about who should do some task or another, but
never quarrels. There is real co-operation, based on a distribution
of tasks in play.

Such play is occasionally associated with a 'house' built of mud
by a brother who often does not share the play, or actually uses
the 'house' for his cattle play while his sisters are 'cooking' nearby.
It is said, also, that housekeeping play sometimes branches into
sexual play, little boys and girls pretending to be husband and
wife and trying to copulate. Detailed inquiry shows that this is
not common. The usual method of sexual experimentation at
this stage of development follows the pattern of adolescence. Small
boys 'woo sweethearts' with little gifts, and sexual experiments
occur in connection with dancing or by chance opportunities.[1]

[1] See Fortes (1936b), loc. cit.

Girls' housekeeping play ceases with the beginning of pubescence. Not only are they by then already taking a full share in real housekeeping but their interests are turning to youths and marriage. Housekeeping play, in which the child rehearses the interests which it has as yet neither the skill nor the degree of social development to satisfy, and expresses the wishes which pubescence brings near to realization, has served its purpose.

The corresponding play of boys at this stage is 'cattle-keeping'. It occurs more sporadically than the girls' play, particularly with older boys; but as much fantasy and invention go into it. Sometimes, indeed, it seems like an overt day-dream of leisure hours. One boy alone, or two, usually play. When the group is bigger more boisterous or recreational play ousts it – they wrestle, shoot arrows, gamble with ground-nuts, or at certain seasons pitch bangles and hoops of plaited reed. Boys of all ages are far less placid than girls.

The 'cattle' play involves constructing a 'house' in which the 'gateway' and the 'stable' are prominently indicated and finding something to serve as 'cattle' – sheep and goats never seem to enter. These activities, and moving the cattle about in and out of the 'stable', with murmured remarks in monologue or addressed to a companion, constitutes the play. I have never been able to record this accompanying speech as I have always had to watch cattle play from a distance. Boys say that it is 'about bulls and cows'.

'Cattle' play has more of pure fantasy and less reproduction of real activities than 'housekeeping' play. Sometimes the 'house' and 'stable' are built of mud – a circular 'wall', 9 or 10 inches in diameter and 2 inches high, with a space left for the 'gateway', and a smaller circle of mud adjoining it as the 'stable' – to last for a whole dry season. Often it is constructed *ad hoc* – two circles drawn in the sand or made of heaped-up sand may be enough. A great variety of objects serve as 'cattle'. Boys who like modelling may make clay figures of cows and bulls for their play; bits of sticks, leaves of a common shrub which can be opened so as to stand up on their edges, pebbles, and other things are used. Yet despite the meagre materials and the paucity of mimetic content the play fascinates boys from the age of 5 or 6, when they are still too young to go out with the herds, to about 12, when they may be full-blown herd-boys. Even if their

1*a*. A young nursemaid, amusing herself and the baby with a *Kinkayax* rattle.

1*b*. Housekeeping play. One of the girls combines the duty of looking after an infant with her play. Note the small boy in the party, and the 'utensils'.

2*a*. A small boy 'sacrificing' to his own 'shrine' (posed).

2*b*. Cleaning groundnuts, one of the ways of learning to use a hoe. The child on the extreme right is a girl.

3*a*. Recreational play: *kuobon* (posed)

3*b*. A wrestling match—with an eye on the goat lest it wander into the growing crops.

4a. Building play. In the background a newly-built room. On the right of the group a big boy demonstrating to a small boy. Behind them part of a 'house' they have finished. On the left, girls making the swish and explaining the process to the ethnographer's wife.

4b. Learning marksmanship. Note how intently the two smaller boys, who have only recently acquired real bows, are watching the technique of their older companion.

families have no cattle and they have never followed a herd regularly, they play at it. The interest and identification are active nevertheless.

At this stage, too, domestic ritual begins to be reflected in the play of boys. A boy's or girl's schema of ritual and religious ideology at the age of 9 or 10 includes the main structural principles of the system. As his knowledge has been acquired by attending at sacrifices, he knows most about the ritual acts and conventional formulas connected with sacrifice and least about the beliefs and theories. He is familiar with all the concepts of Tale religion and magic but cannot assign them accurately to their relevant context. He knows that ancestor spirits and medicines are different, and can even describe some of the latter by their functions, but cannot elucidate these differences. He knows also and believes that health, prosperity and success depend on mystical agencies, that sickness, death, and misfortune are caused by them, and that sacrifices must be made to placate them or to expiate offences. He has heard talk of all this and seen consultations of diviners. As an infant, perhaps, he has been called by his mother to get off a partition wall 'lest the spirits push you off', or has seen food put out 'for the spirits' during the ritual festivals. He knows what different types of shrines look like and what are their appurtenances. But it is surface knowledge, confused in details and full of gaps.

Ritual is men's business, though women are well versed in it. Hence it emerges mostly in the play of boys and not of girls. Significantly, it is permitted till pubescence, that is, as long as a boy is not likely to take a responsible part in real ritual. After that he is liable to have to accept an ancestor spirit demanding real sacrifices and may no longer play at it. At 13 or 14 years of age, when a boy's ritual schemas approximate those of an adult, he fully understands and acquiesces in this prohibition. It suggests however, that playing at ritual has a different value for children than actual ritual has for adults. Children share the adults' interest in ritual and accept its prescriptions, but not the adults' emotional relationship to ritual. In their play they express this interest and their identifications, rehearse their knowledge, and integrate it with the rest of their educational achievements.

Small boys build shrines a few inches high for themselves in a corner of the cattle-yard. They take great pains to achieve verisimilitude and neatness, and their inventiveness is remarkable. *Kalɔmpoo* husks are turned into medicine-pots; fragments of calabash or potsherd represent the hoe-blade which is essential to many ancestor shrines; a pronged twig is a shrine's 'tree'; the tail of a stillborn kid or lamb, tied with string and feathers in the same way as adults do, is a shrine's 'tail', another object commonly dedicated to real shrines. There are 'roots' – of grass – as in adult medicine-pots, and other appurtenances. Whenever they build miniature houses, during the building season, 'shrines' are added and, as in real life, each has his own.

Play with these shrines is woven into other play activities, and it revolves around sacrifice. When a small boy goes out hunting for fieldmice or birds, if he happens to have a 'shrine' he will 'give it water', i.e. pour a libation to it. Ashes represent flour, which is stirred up in water as in a real sacrifice. He invokes the shrine, 'My father' (but never mentioning names as in real sacrifice since his own father is probably still alive), 'accept this water and grant that I have successful hunting. If I kill an animal, I will give you a dog.' Some time later he may catch a live mouse, and when he has played with it to satiety he 'sacrifices' it on his shrine – this is the promised 'dog'. A nestling bird found alive is 'sacrificed' as a 'fowl' or 'guinea-fowl'. If he finds a live mouse or bird by chance it will always be taken home and 'sacrificed' thus before it is cut up and eaten. Taboos like those of adults are invented for his shrines. Fetishes, like those of adults, accept only red and black 'fowls'; other shrines only white ones. Siikaɔni, a small boy of about 7 or 8, built himself a *loo* fetish, which can be dispatched to 'tie up' any one who might interfere with one's enterprises. Siikaɔni pretended to use his *loo* to keep the parents of his 'sweetheart' out of the way when he went to see her – a frequent use of a real *loo* by young men.

Of the imaginative and constructive play produced as a response to current social activities in which children co-operate, I shall instance only building. During the building season a favourite preoccupation of the children is to build miniature rooms or houses. It is a co-operative group enterprise, boys and girls frequently working together, led by a boy of about 12–13. It is carried on in the intervals between helping the men. Sometimes they are content to build only walls, a few inches high, but when the leader is keen they undertake a complete replica of a house – walls a foot or more high, several rooms properly arranged, a roof, a beaten inner court, and shrines. The work goes on for days on end and needs planning and organization. The girls and smaller boys make the swish and roll it into pellets, the older boys do the actual building, often with extraordinary skill. When the 'house' is built and roofed and the walls have dried, if the enthusiasm for building still lasts, the girls plaster the walls and beat the inner court in the way they have learnt by assisting their mothers with these tasks in connection with the real house.

Children's arts and crafts have a play value in that they are practised purely for pleasure and have a seasonal incidence. But they demand considerable skill of eye and hand, and individual differences in ability are noticeable. Towards the end of the rainy season a strong and supple reed springs up in profusion along the watercourses. Young people and children pluck these reeds to plait bangles, necklets, small panels to hang over the chest as decorations, and waistbands. These things are worn by young and old at the festival dances. But whereas the young men and women plait sporadically a few things for themselves, their sweetheart, or a child, children do so continuously and absorbedly. From July to September one sees them sitting about or strolling about in small groups, plaiting reed decorations. The boys have a game played by tossing or bowling the reed bangles and waistbands at a mark. A group plays, and the winner collects all. Girls of all ages love decking themselves from head to foot in this reedware. The technique of plaiting these articles, which is the same as that employed in the manufacture of a number of utilitarian objects, is fairly elaborate. It is gradually learnt between the ages of about 8 and 11.

Less widely practised is the art of modelling clay figures. Girls sometimes model, but it is chiefly a diversion of boys. Cattle and other animals, humans, horses accoutred and with riders astride are the usual subjects. A taste for modelling appears to depend on talent to a great extent. Many boys never acquire the art; others take such delight in it that they devote the whole dry season to producing dozens of clay figures for themselves, brothers and sisters, and friends. Gifted boys model extremely cleverly. I knew two boys of 12 or 13, sons of a chief, who made clay horses and manufactured saddlery and trappings of old rags and bells and ornaments of pieces of tin to adorn the figures. Older boys or youths teach small boys by correcting errors they make in modelling; but I have never found a boy of under 8 or 9 years able to model well.

Play from pubescence to adolescence. The last stage of childhood coincides with the rapid absorption of the child into the economic system and his or her gradual acquisition of a responsible status in the social structure. By the age of 14 or 15 most girls are already married or being courted in marriage. They take their household duties more lightly perhaps than older women with children, but their childhood education is complete. Their education in the duties and responsibilities of wifehood and motherhood lies outside the scope of this paper. 'To play' now means to join in the dance or to dress up and go to market, there to gossip and flirt. These are recreations merely, like conversation in the evening after a good meal, when the whole family sits or sprawls about in one of the inner courts or in front of the gateway. Such 'play' is educative in quite a different sense to that of childhood.

Boys, too, between the ages of 12 or 13 and 16 to 18 are at the stage of transition from childhood to young manhood. The imaginative play still prominent at the beginning of this period is given up by degrees and usually altogether abandoned when puberty is established. Like the adolescent girl, the boy of 16 or so finds his principal recreation in the dance at certain seasons. He, too, begins to frequent markets when time permits, for he is greatly preoccupied with the opposite sex, with courtship and flirtations and even transient love affairs, and there is no place like the market for pretty girls. An adolescent youth is already applying the deftness and skill acquired in juvenile play or in the arts and crafts of his boyhood to practical ends. A 16-year-old takes an active part in building and thatching and in the manufacture of bows and arrows, or in the practice of crafts like leatherwork or the forging of tools and implements.

The transition from boyhood to manhood can readily be observed in the development of farming interests and skill during this period. The boy of 10 to 12 is extremely keen to plant, hoe and weed. Helping his father, he sows ground-nuts for himself amongst his father's early millet. Frequently he has a small plot of cereals, a few yards square, in a useless corner of one of his father's fields. He hoes and plants and weeds his plot with great energy and zest, though somewhat crudely, borrowing one of his father's discarded hoes for this purpose. He assists his mother to farm her ground-nuts and beans. But his efforts make no difference to the family commissariat or to the care and sustenance given him by his parents. He is still experimenting without responsibility, though with great earnestness. Two or three years later the play element has vanished. If he cultivates a personal plot he makes an effort to beg land which is agriculturally good, and works with the avowed purpose of obtaining a crop which, though minute compared with the needs of the family, suffices to buy himself a cap or a loin-cloth. The time he can now devote to his own plot or to his own ground-nuts must be adjusted in accordance with his responsibilities as a contributor to the family economy.

With boys, therefore, as with girls, the completion of their childhood education marks the end of childhood play. Mimetic and imaginative experimentation becomes redundant when the individual attains social responsibility and maturity. The play of Tale children changes, as we have seen, *pari passu* with their advancing maturity, contributing at each stage to the elaboration and integration of those interests, skills, and observances the mastery and acceptance of which is the final result of their education.

VI. SYNOPTIC CHART OF EDUCATIONAL DEVELOPMENT

In conclusion, and in order to bring the preceding analysis of the education of Tale children to a focus, I append a synoptic chart setting out the main trends of the development of their interests and skills as these are reflected in their economic duties and activities, on the one hand, and in their play on the other. The reader will realize that these norms are necessarily crude and approximate. I have not ventured to include observances in the chart, since the trend of their acquisition cannot easily be analysed into approximate stages.

CHART OF DEVELOPMENT

BOYS

Economic Duties and Activities   *Play*

3–6 years

None at first. Towards end of this period begin to assist in pegging out goats; scaring birds from newly sown fields and from crops; accompany family sowing and harvesting parties; using hoe in quasi-play to glean ground-nuts in company of older siblings.

Exuberant motor and exploratory play. Use mimetic toys (bow, drum, etc.) in egocentric play. Towards end of period social and imaginative play with 'cattle' and 'house-building' commences, often in company of older children of either sex, as well as recreational games and dancing.

6–9 years

These duties now fully established. Help in house-building by carrying swish. Assist in sowing and harvesting. Towards end of period begin to go out with the herd-boys, and to care for poultry.

Imaginative 'cattle' and 'house-building' play common, the latter often reflecting current economic activity of adults. Practice with bow and arrow in marksmanship competitions, and 'hunting' with groups of comrades begun. Recreational games and dancing established. Modelling clay figures and plaiting begun. Ritual play begun.

9–12 years

Fully responsible cattle-herding. Care for poultry. Assisting parents in hoeing and care of crops, but without responsibility. Farming own small plots and ground-nuts but in quasi-play. Sons of specialist craftsmen assist fathers in subsidiary capacity – 'learning by looking'.

Further development of preceding forms of play, especially of ritual play. Claymodelling and plaiting established. Recreational games and dancing more skilful. Quasi-play farming.

Sexual dichotomy in work and play established.

12–15 years

Duties as in preceding period but more responsible. Responsible care of poultry, sometimes own property. Leaders of herd-boys. Real farming of own plots and in co-operation with older members of family established by end of period. Sons of specialists experimentally making things.

Imaginative play abandoned. Dancing the principal recreation. Ritual play abandoned. Modelling gradually abandoned. Plaiting for personal decoration mainly. Regular sweethearting commences.

GIRLS

Economic Duties and Activities   *Play*

3–6 years

None at first. Towards end of period the same duties as small boys. Frequent nursing of infants. Accompany mothers to water-hole and begin to carry tiny water-pots. Help in simple domestic tasks such as sweeping.

Exuberant motor and exploratory play. Attached to older sisters and drawn into their 'housekeeping' play. Towards end of period begin to take active social part in the latter and begin recreational play and dancing. often found in mixed sex groups.

6–9 years

Duties of previous period established. Responsible co-operation in water-carrying and simpler domestic duties. Help in cooking and in activities associated with food-preparation, such as searching for wild edible herbs. Accompany family parties at sowing and harvesting, giving quasi-playful help. Carry swish at building operations and assist women in plastering and floor-beating, but still with a play element.

'Housekeeping' play usual. Recreational play and dancing established. Begin to learn plaiting. Participate in 'building' play of boys, mimicking current women's activities, e.g. plastering.

9–12 years

All domestic duties can be entrusted to them by end of this period – water-carrying, cooking, care of infants, etc. Assisting in building and plastering, etc., more responsibly. Often sent to market to buy and sell. Help in women's part of the work at sowing and harvest times.

'Housekeeping' play continues, gradually fading out at end of this period. Dancing becomes principal recreation. Plaiting both for decoration and use established. Begin to have sweethearts but not yet with serious intent.

Sexual dichotomy in work and play established.

12–15 years

Responsible part in all domestic duties of everyday life, and of those associated with ceremonial occasions. Go for firewood and collect shea-fruits in the bush, and help to prepare shea butter. (Marriage a very near prospect.)

Note. Care of infants and children is a duty of girls at all ages. Boys also are frequently entrusted with this task.

Imaginative play abandoned. Dancing the main recreation. Courtship and hetero-sexual interests occupy a great deal of time and attention. Actively participate in the social side of funeral ceremonies, etc., in the role of marriageable girls.

9

Radcliffe-Brown's Contributions to the Study of Social Organization[1]

It is astonishing to conclude from some of the reviews of this selection of Professor Radcliffe-Brown's essays and papers (1952) that there are professional sociologists who owe their first acquaintance with his work to this publication. For it can be categorically asserted that each of the papers here reprinted has been a major landmark in the growth of social anthropology over the thirty years since the publication in 1924 of the famous 'Mother's Brother' paper. They embody a series of discoveries and hypotheses which changed the course of anthropological study, at any rate in Great Britain, and have only been fully appreciated by anthropologists in the past decade or so. Readers of American anthropological journals[2] will find that some of the most influential anthropologists there have recently isolated a contemporary 'British', or as some prefer to say 'structuralist', school of social anthropology. The measure of aptness there is in this label refers to a frame of analysis that has grown primarily from the assimilation of Radcliffe-Brown's ideas and theories into the body of British anthropological research. This assimilation is so complete that one often repeats Radcliffe-Brown's arguments, even his examples, without realizing it. How often have I, and other social anthropologists, not used the example of a Court of Law to illustrate the assumption that order and consistency in its institutions are necessary for the social life of a community to go on smoothly? But we seldom recollect that it was first used to illustrate this very argument by Radcliffe-Brown in his

[1] Reprinted from *The British Journal of Sociology*, **6**, No. 1, March 1955.
[2] *Cf.* the papers by Murdock (1951) and by Eggan (1954).

paper on 'Methods of Ethnology and Social Anthropology' (1923).

I think it is a pity that this paper is not reprinted here, for it is the first published statement, in a generalized form, of what Radcliffe-Brown considers to be the distinctive methods, theories and field of inquiry of social anthropology. It is true that the essentials of his point of view are to be found in the *Andaman Islanders*, which appeared at the same time (1922), but the reader has to extract them and generalize them for himself and this is not a task that appeals to the non-anthropologist. It is true, also, that the thesis of the 1923 paper is expanded and is more incisively stated in the 1931 Presidential Address to Section H of the *British Association for the Advancement of Science*[1] and in the 1935 paper 'On the Concept of Function'. Historically, however, the 1923 paper is the most significant one, for by 1931 the 'functionalist' conception had already been translated into field ethnography and theoretical controversy by Malinowski and his pupils in this country and by Radcliffe-Brown and his pupils in South Africa and Australia. Evans-Pritchard, Firth, Hogbin, Audrey Richards and Schapera had already published the first fruits of their 'functionalist' field work, and Mrs Hoernlé was teaching the theory and the method in South Africa. It had, indeed, also made such redoubtable American converts as Margaret Mead[2] and (though she apparently did not admit this), Ruth Benedict.[3] And furthermore, the divergence of theoretical emphasis between Malinowski's functionalism and Radcliffe-Brown's was then already apparent, whereas in 1923 it was only foreshadowed, and they stood together in opposition to the most influential tendencies of the day. Both insisted, in their diverse ways, on the autonomy of social anthropology as an 'inductive science' as against the historical character and methods of ethnology on the one hand and the interest in the thoughts, feelings and actions of individuals, which is the province, strictly speaking, of psychology, on the other.

[1] 'The present position of anthropological studies', British Assoc. for the Advancement of Science, *Annual Report*, 1931, p. 141.

[2] See her *Kinship in the Admiralty Islands*, American Museum of Natural History, 1934.

[3] This is documented, to my mind convincingly, by Leslie White in his paper 'Evolutionary Stages, Progress and the Evaluation of Cultures', *S.W.J.A.*, **3**, No. 3, 1947, pp. 184-5. But it had previously been remarked by Radcliffe-Brown himself in his paper on the 'Concept of Function', when it was originally published in 1935 in *The American Anthropologist*.

This was the point emphatically put forward in Radcliffe-Brown's 1923 paper and reiterated in the 1931 and 1935 papers. It is now taken for granted by many social anthropologists.

What Radcliffe-Brown did, with that inimitable lucidity and precision for which comparison with the literary habits of present-day social scientists make us ever grateful, was to distinguish between the search for (unverifiable) historical and evolutionary origins – the preoccupation of ethnology – and the study of the (verifiable) laws of custom and social organization – the province of social anthropology. And this methodological distinction was driven home with two or three telling examples. It is these examples which were later expanded into the analytical papers collected in *Structure and Function in Primitive Society* (1952), and it is from these papers, each dealing with a problem in the analysis of ethnographic data, rather than from the 'programme and methods' papers, that British social anthropology of today derives so much of its basic frame of thought. Those social scientists other than anthropologists who know Radcliffe-Brown's work, tend to pay attention only to his methodological articles; but the vital theoretical contribution lies in the ethnographically centred papers. It is in them that Radcliffe-Brown's adherence to the view that there are laws of social life which can be established by systematic and empirical investigation, is brilliantly justified.

These papers are mainly concerned with problems of social organization, or as Radcliffe-Brown now prefers to call it, social structure, as he expounded this concept in the 1940a paper. In this department of anthropological study, Radcliffe-Brown has enunciated general principles to which it is surely quibbling to refuse the name of laws. No better tribute to the laws of kinship structure discovered by Radcliffe-Brown and stated in the 1941 paper could be possible than that tacitly rendered by Professor R. H. Lowie in his (1949) book *Social Organization*. For Lowie's first major work in this field, *Primitive Society*, was published in 1922, and is warmly referred to in Radcliffe-Brown's paper of 1923. But at that time Lowie was still non-committal, if not definitely sceptical about the possibility of 'laws' of human social organization, and his main concern was to demonstrate the inadequacy of practically all existing 'historical' theories accounting for kinship customs and institutions in the field of social and political organization. In *Social Organization* he accepts Radcliffe-Brown's two principles of the 'Unity of the sibling group' and 'The Unity of the lineage' and agrees that the former explains the levirate and the sororate. This, from Lowie, is an enormous tribute. For it admits that these customs show the workings of general laws of social organization.

To understand the revolutionary significance of the papers published by Radcliffe-Brown between 1923 and 1935 we must look at the way in which the same problems were being tackled by the leading anthropologists of the day. In the study of primitive kinship and social organization there was, till his death in 1922, no one in England of authority equal to that of W. H. R. Rivers. In 1921 he gave a presidential address to the Royal Anthropological Institute.[1] 'It is a characteristic of the simpler societies ...' he said, 'that, though it is possible to distinguish in their cultures the different aspects we label social, political (etc.) these aspects are so interdependent ... that it is hopeless to expect to understand any one department of culture without an extensive study of other departments....'

These words might have been used by Radcliffe-Brown or Malinowski. But contrary to the conclusion *they* would have drawn, what Rivers infers is the need for deeper studies of the 'history' of primitive cultures and the hypothesis which he recommends without reservation is that of Elliot Smith and Perry. His theory is that just because they are so remote and off the track of subsequent cultural influences, the islands of Oceania have preserved customs and institutions of the 'megalithic' civilizations which spread all over the world from ancient Egypt.

This was fully in accord with the approach to problems of social organization he had developed at elaborate length and with painful perversity in *The History of Melanesian Society* (1914). Thus, the then most commonly accepted explanation of the peculiar privileges of a sister's son in relation to his mother's brother, which had been reported from a number of societies with patrilineal descent systems, was that the custom was a survival from an earlier stage of mother-right. Rivers, acquiescing in this theory, put forward another theory as well. He contended that the nomenclature used indicated the origin of this custom in some antecedent rule of marriage, e.g. one by which old men

[1] 'The Unity of Anthropology', *J.R.A.I.*, **52**, 1922, pp. 12 ff.

were obliged to hand over their wives to their sister's sons. Or, to take another curious custom, it is very common for grand-parents and grandchildren to be classed together in the classi-ficatory kinship terminologies of primitive societies. They have a relationship of familiarity and equality, not unlike that of siblings, by contrast with the respect relationship of successive generations. Radcliffe-Brown (1941) shows that this merging, or identification, of alternate generations fits in with the widespread concept of the completion of the cycle of the generations in three successive steps, and with the contrasting roles of grandparent and parent in family systems. It is also a necessary feature of kinship systems in which 'dual organization' occurs. Rivers, however, seeks an explanation by assuming an 'ancient rule' by which a man married his brother's grand-daughter. To account for the possi-bility of such rules Rivers is 'driven to assume a state of society in which these elders had in some way acquired ... a position ... able to monopolize all the young women' (1914, vol. 2, p. 59). We can laugh at this kind of nonsense now; in the 1920's it was no more than a special case of the generally prevailing climate of thought in the anthropological sciences.

In America, Kroeber, just reaching the peak of his high repu-tation, was less doctrinaire, but he was also much less rigorous than Rivers, and his theoretical aims were of the same general kind. Indeed as late as 1934 he still seemed to hold to his view that kinship terminologies were primarily linguistic phenomena to be understood in terms of linguistic history and psychology and having only a somewhat loose connection with social institutions. Criticism by Radcliffe-Brown led him to change his views and bring them nearer to Radcliffe-Brown's. But Kroeber has always maintained that he is an historian, and evocation is more in his line than analysis.[1]

The American scholar who still comes nearest to Radcliffe-Brown in interests is R. H. Lowie, whose reluctance to accept generalizations in anthropology I have already mentioned. His main interest has been the classification of primitive customs and institutions with a view to more accurate histories of their development than the evolutionists were able to provide, rather than analysis leading to statements of general tendencies which

[1] See the collected papers of A. L. Kroeber, *The Nature of Culture*, Univ. of Chicago Press, 1953, chs. 23 and 24.

Radcliffe-Brown has pursued. Even Malinowski's novel approach to kinship and social organization raises sociological difficulties when closely looked at. His 'biographical' method of tracing the 'extension' of kinship attitudes and sentiments from the kinsfolk in the parental family to their collaterals and so outward to wider kin groupings, focuses attention on the conditioning process and on the ostensible emotional elements in kinship. On this basis, all conventions of grouping kin outside the elementary family are reduced to the same elements and it is impossible to understand the kind of observed differences embodied in variations in terminology. In fact all these approaches miss the sociological essentials. But Radcliffe-Brown saw them early and developed them in a series of papers that began with the publication of his first Australian field work in 1913[1] and culminates in the masterly paper of 1940.

The essence of the hypothesis was stated in 1923. Kinship customs and usages form a system. The relations of mother's brother and sister's son among the Bathonga cannot be understood without taking into account the correlative relationship of father's sister and brother's son, as well as the fact that this is a society segmented into groups in which descent and succession pass in the patrilineal line. The peculiar usage by which a Thonga calls his mother's brother 'male mother' behaves towards him with familiarity, and expects from him indulgence and friendly help, whereas his 'behaviour pattern' to his father's brother and sister is one of respect and expectation of severity and formality fall into place when seen not as isolated customs but as parts of a system. It is not a question of extending outwards, by the process of psychological conditioning, the emotional attitudes and sentiments implanted in infancy by parents, but of what, in the later development of his theory Radcliffe-Brown referred to as norms of behaviour and jural principles; and the key to the mode of extension is the principle of the equivalence of brothers, which by 1940 becomes the more general and fundamental principle of the unity of the sibling group. The individual tends to be merged in a group and so behaviour originating in the relationship to one member of a group – here the group of brothers and sisters – gets applied to all of them.

It should be emphasized that the analysis is strictly sociological

[1] 'Three Tribes of Western Australia', *J.R.A.I.*, **43**, 1913.

and becomes more and more so with each successive refinement of the basic hypothesis. In the 1940 paper and much more so in the more detailed statement of the theory in the Introduction to *African Systems of Kinship and Marriage* (1950) the argument is wholly purified of psychological implications in the sense of hidden assumptions about a conditioning process working on individuals. The principle of the unity of the sibling group, and its correlative the principle of the unity of the lineage, are enunciated and handled as sociological principles. They refer to the jural status of persons from the point of view of the *world at large*, not to emotional identifications. Thus the key to classificatory systems of kinship is the principle that collateral kin and lineal kin are identified by the mechanism of sibling unity; and the related distribution of rights and duties, the patterns of conduct and of sentiment, what are often designated the forms of kinship behaviour found in a society, are perpetuated by the principle of lineage unity. No other hypothesis has yet been put forward which accounts so economically and with such generality for the structure and mode of operation of classificatory kinship systems.

As I have already mentioned, Lowie now concedes the validity of these principles. In fact they are statements of laws of kinship organization which have a validity of the same order as the statements of general tendencies enshrined in such economic concepts as that of marginal utility or such psychological concepts as that of the conditioned reflex. They belong to a cluster of interconnected propositions of varying degrees of generality concerning kinship and social organization set out in the 1940 papers on 'Kinship' and on 'Social Structure' and in the 1950 'Introduction'.

These propositions have been verified by field observations many times in the past decade. They have opened up new problems for research; above all they provided by far the most fruitful hypotheses to account for the structure and mode of operation of classificatory kinship systems. And this is due to Radcliffe-Brown's theoretical clarity in dealing with kinship on strictly sociological lines, which means differentiating the *jural* significance of kinship from its *psychological meaning*.

Most of the futile controversies about kinship turn on failure to make this distinction. A crucial instance is provided by Malinowski.[1] It is a striking illustration of the difference between his method of approach, referred to above, and that developed by Radcliffe-Brown. On p. 447 of the *Sexual Life of Savages*, Malinowski records faithfully the terminology by which the father's sister's son is called by the term for '*father*', among the Trobrianders. He is so preoccupied with analysing the conditioning of the emotional attitudes about incest which he strives to relate to the extension of the kinship terms, that this usage puzzles him. He calls it 'anomalous'; and he falls back on the irrelevant argument – incidentally, an echo of Rivers, whose approach to kinship he never stopped castigating – that this is an example of the influence of language on custom! Yet the usage is crystal clear, and perfectly consistent with Trobriand jural institutions, if it is analysed by Radcliffe-Brown's methods. It is like any other 'Choctaw' system, as described by Radcliffe-Brown (1940). It is a common custom in all such systems, in which inheritance and succession follow the maternal line, for the class of a man's potential heirs and successors (i.e. his sister's sons) to be identified in terminology and social position with the father by his children. For, to put it at its simplest, a sister's son will one day take over their father's jural status and roles in relation to themselves and to society at large; or, in more abstract terms, the rights vested in the father's descent line, on the principle of the unity of the lineage, vest in his heir too, *inter alia* as a representative of that line. Had Malinowski perceived the distinction between the jural and legal significance of *kinship systems* on the one hand, and the psychological elements in the *person-to-person conduct* of relatives by birth and marriage to one another on the other, he would have avoided this blunder.

This basic distinction between the *jural* aspects and functions of a kinship system and the *affective*, or if we prefer, psychological meanings of the customs and usages in which the system comes to expression, has been more difficult to establish than might be thought. It involves a confusion of the same type as that of the 'history versus science' controversy in anthropology. And as with the latter issue no living anthropologist has done so much to bring discipline and clarity into our thinking as has Radcliffe-Brown.

An outstanding contribution in this connection is the paper on

[1] For a more detailed discussion of this point, see my paper 'The Structure of Unilineal Descent Groups', *A.A.*, 55, 1, 1953, pp. 17–41.

'Patrilineal and Matrilineal Succession' (1935a). This is the paper in which he expounds the concept of 'perpetual corporate succession', which derives from Maine, in its ethnographic application. What he demonstrates is that 'matriliny' and 'patriliny' are simply alternative ways of reaching a solution of the fundamental problem of succession, which is to determine unequivocally where rights over persons reside and to stabilize these rights. These rights must be unequivocal and consistent over time as well as at a given time, hence the laws of succession and inheritance found in any society are related to the more general problems of maintaining functional consistency amongst the constituent parts of the social system and ensuring its stability and continuity over time. The hypotheses and the concepts developed in this paper have been the starting point of many of the recent ethnographical and theoretical writings of British social anthropologists which are regarded in America as so distinctive of British 'structuralism'.

A major development in this work has been the analysis of primitive political institutions, especially in Africa; and a key principle has been Radcliffe-Brown's concept of descent as fundamentally a jural or legal notion that *inter alia* regulates the forms of grouping in which corporate 'ownership' of citizen's rights, land, property, as well as political and religious office and control over the reproductive resources, is vested. It is such corporate descent groups, or lineages, which in some African societies constitute the only 'sovereign' groups, and in others, the basic 'constituencies' which provide a chief's councillors and military leaders. In particular, this way of looking at social structure has revealed the quite complex machinery of political and legal organization in societies formerly dismissed as being simply devoid of political or legal institutions. All this is admirably expounded in Radcliffe-Brown's 'Preface' to *African Political Systems*.[1] It has thrown new light on such controversial subjects as the nature of the feud, now seen to be a regular concomitant of segmentation by descent groups in societies lacking central government, especially if they are subject to relatively harsh environmental or external political pressures, and it has elucidated such other classical topics of anthropological controversy as the rules of preferential and prohibited marriage.

[1] See M. Fortes and E. E. Evans-Pritchard (eds.), 1940.

In the latter case we are confronted, firstly with the effect of descent rules which identify as equivalent or reciprocal *legal persons* those who are forbidden to marry. But we also observe the effect of the principle of functional consistency which has been the cornerstone of Radcliffe-Brown's work for thirty years. This principle, which Radcliffe-Brown derived from Durkheim, is the central theme of all his purely theoretical papers. It is expounded with unerring logic in the paper, 'On the Concept of Function in Social Science'. Though this paper was first published in the *American Anthropologist* in 1935, its reverberations are still being felt both among American and among Cis-Atlantic anthropologists.

As he states the problem in his latest exposition of his theory of kinship[1] both prohibited and preferential marriages must be analysed with reference to the effect they have in preserving and maintaining an existing kinship structure. A prohibited marriage is one which threatens to disrupt it; a preferential marriage is one which is calculated to maintain it.

It has been said that this mode of reasoning is no more than 'mere tautologies and platitudes on the level of common-sense deduction' (Evans-Pritchard 1951b, p. 57). Yet no one has provided such impressive ethnographic vindication of this principle as the author of this statement (who, as it happens, is joint author with Professor F. Eggan of the foreword to the present book) in his admirable studies of Nuer social organization (Evans-Pritchard 1940a, 1951a). The ostensible vagaries of Nuer marriage laws which enjoin ghost-marriage (as Professor Evans-Pritchard graphically labels the custom, also found among some East and South African Bantu tribes, of marrying a wife 'to the name' of a dead man) and the levirate, permit women to take wives as if they were men, and tolerate certain forms of concubinage, fall into order if they are looked at in the light of this principle. For whatever arrangements about sexual and domestic services they involve, none of them violates the basic rule of the primacy of patrilineal descent, in the jural sense, in fixing lineage membership for legitimate offspring, or runs counter to the rule of the unity of the sibling group as this is expressed in Nuer custom. What from the descriptive point of view, may look like latitude verging on licence, is perfectly orderly in terms of the

[1] In the Introduction to *African Systems of Kinship and Marriage*, pp. 60 ff.

hypothesis that a form of *marriage* (which is not necessarily the same thing as licit sexual relations) is not allowed if it threatens to disrupt the kinship system as a whole. That is why, among the Nuer as among very many African tribes, the status of begetter (*genitor*) is precisely distinguished from that of legal father (*pater*).[1] The hypothesis of consistency is therefore immediately important for the procedure of investigation it entails; and this is best illustrated in Radcliffe-Brown's studies of ritual subjects, as in the papers on Taboo (his Frazer Lecture, 1939), on Totemism (1929), and on Religion and Society (his Henry Myers Lecture, 1945). But before I deal with them there is a criticism of Radcliffe-Brown's approach to society which needs to be considered.

It is thus stated in one review of the present book: 'It is the human quality of resistance to the letter of the law that seems to be left out of many of Mr Radcliffe-Brown's calculations' (*The Economist*, September 6, 1952). There is a point in this. Radcliffe-Brown's concept of social structure emphasizes the norms of social behaviour and the institutions for maintaining them. This calls for a consideration of the nature of law and morals. His views are summed up in the articles on 'Social Sanctions' and 'Primitive Law' from the *Encyclopædia of the Social Sciences* (1933). They reflect his morphological bias in that they are concerned with definition and classification of primitive juridical institutions not with analyses of how they work. Thus he says of sanctions: 'The sanctions existing in a community constitute motives in the individual for the regulation of his conduct in conformity with usage'; and again, 'What is called conscience is thus in the widest sense the reflex in the individual of the sanctions of the society.' Good. But these unexceptionable statements have nowhere (since the *Andaman Islanders* which followed Durkheim fairly closely) been translated into analysis of mechanism and function. We do not know what Radcliffe-Brown means by 'reflex' in this context, or how he conceives of the process of what is now often called – by a term borrowed from psychoanalysis – 'internalization'. In fact, all he says is that there are correspondences between institutional norms and arrangements, and patterns of individual motivation. But he does not inquire further as to how the 'individual' as distinct from the 'person', to use the terminology

[1] These terms were introduced by Radcliffe-Brown in his famous monograph, 'The Social Organization of Australian Tribes', *Oceania*, I, 1930–1.

he uses in his 1950 'Introduction', acts, feels, judges and chooses as a member of his society.

If those who criticize Radcliffe-Brown for leaving the individual out of his considerations really understood what he is aiming at they would see that what they are reproaching him for doing is what he explicitly sets out to do. It is related that when A. C. Haddon read a book by a distinguished lady anthropologist, he remarked: 'Humph, when I went to the Torres Straits I took along three psychologists, but I see I would have done better to have taken along a lady novelist.' Apocryphal or not, this story sums up the kind of thing many of Radcliffe-Brown's critics expect social anthropology to be about and naturally find lacking in his work. There is room, perhaps even need, for literary virtuosity and for obtrusion of the writer's ego in the descriptive ethnography to which the late Dr Marett gave the name of 'the higher gossipry'. There may be place in this branch of anthropological writing for the imputation of motives and the dramatization of actions. But Radcliffe-Brown is not writing particular ethnographies. He is not even trying to include in his range of analysis everything that goes on in social life. What his aim has consistently been is to establish generalizations of a limited kind about the 'network of social relations' that is the fundamental reality of social life everywhere; and no one will dispute his reiterated contention[1] that this can only be done by using some kind of comparative method.

This objective focuses attention on the regularities of usage and conduct in social relations and on those general characteristics that transcend particular times, places and bodies of custom. This aim and method are brilliantly illustrated in the two papers on Joking Relationships (1940b, 1949). One way of demonstrating the inter-dependence of kinship relations in a society is to classify them in accordance with the norms of conduct – what Radcliffe-Brown formerly called standardized patterns of behaviour – required or permitted in different relationships. The regularity – the standardization, if we like – of these norms is a necessary feature of the regularity of the relationships. They are matters of right and duty, as well as of standardized sentiment. Thus it is found that children are commonly, probably universally,

[1] As for instance in his Huxley Memorial Lecture. 'The Comparative Method in Social Anthropology', *J.R.A.I.*, **81**, 1951.

required to show *respect* in their conduct towards their parents but, by contrast, may and do act with freedom and *familiarity* towards their siblings. In many societies it is a regular and stereotyped practice for certain categories of tribal neighbours, kinsfolk and affines to *tease* or abuse, or take other liberties with each other, that would, in different contexts of social relationship, be mortal offences. In these societies, we also commonly find stereotyped *avoidance* rules for certain kin.

Thus a person's kinsfolk are distributed in at least these four classes: those who must be accorded respect, those who may be treated familiarly, those who must be avoided and those who may be teased in set ways; and these four categories are mutually exclusive, in conformity with the principle of consistency. A theory of joking relationships in general is only possible if they are seen to be part of such a system of ordering all kinship relationships.

The joking rules characteristically occur in 'structural situations' where the parties are in a position of detachment or even opposition in terms of one significant social relationship and in a position of attachment or even friendship in terms of another. The commonest case is that of brothers-in-law in societies with corporate patrilineal descent groups. For as members of different exogamous groups they are detached, sometimes to the point of opposition, and this is a necessary condition for their being brothers-in-law; but as husband and brother, respectively, of the same woman they have a personal bond arising out of their common interest in her. The joking relationship is a way of reconciling these two divergent tendencies and so contributes to keep the whole system of kinship relations in equilibrium.

This is a very bald paraphrase of Radcliffe-Brown's hypothesis; but what is significant for my present argument is that the second paper (1949) is an answer to a criticism of the type I have been discussing. This criticism came from Professor M. Griaule who rejected Radcliffe-Brown's hypothesis as not applicable to the Dogon of the French Sudan. Griaule prefers the 'explanation' offered by the Dogon themselves, that their custom of exchanging insults with a neighbouring tribe, with whom they may not marry or fight, is a way of 'purifying the liver', the seat of disturbed feeling, and that it is founded upon a myth of the twin origin of the two peoples. Vivid and picturesque as this 'explanation' is, it is of limited ethnographic worth only, being no more, indeed, than a traditional Dogon rationalization. It cannot, as Radcliffe-Brown points out, provide an 'explanation' in any theoretical sense that will cover the wide variety of similar joking relationships found even in Africa only, for every tribe has its own traditional 'explanations' and mythology of its joking relationships. What is common to all is not the content of the joking actions and beliefs, nor the ideas and feelings avowed by particular actors, but the 'structural situation', the general character of the social relationships, in which joking is enjoined. This is what Radcliffe-Brown is trying to 'explain' in terms of general principles of social organization; and it is worth adding that though there is plenty of ethnographic description of joking customs in primitive societies, Radcliffe-Brown's is the only attempt to set up a general hypothesis for the institution, and it is an hypothesis that has, so far as it goes, stood every test of field observation up to the present.

'So far as it goes' is fundamental. It is perfectly correct that Radcliffe-Brown's abstractions ignore the realities of the loves and the hates, the fears and desires, the curious beliefs and weird ambitions of uncivilized races, which the lay reader expects all anthropologists to purvey. The obvious answer is that writings of this type may make good literature or historiography; they do not contribute to the building up of a body of scientific generalizations which is Radcliffe-Brown's avowed aim; and this task necessarily entails the isolation of relevant data from the kaleidoscope of actual life. More pertinently, it might then be argued that it is a serious distortion to leave out consideration of motive and, especially, the details of custom and usage as described by ethnographers, in order to focus attention on the structural situation.

This takes us to the heart of current controversy about theory subject-matter and methods in social anthropology.[1] 'To me,' said Dr G. P. Murdock in a paper read at the Fiftieth Annual Meeting of the American Anthropological Association at Chicago, November 15–17, 1951, 'scientific theory in our discipline revolves

---

[1] *Cf.*, for example, G. P. Murdock, 'British Social Anthropology', and Raymond Firth, 'Contemporary British Social Anthropology', *A.A.*, **53**, 1951, pp. 465–89; also F. Eggan 'Social Anthropology and Controlled Comparison', *A.A.*, **56**, 1954, pp. 743–63.

around the construct of culture ...'; the kind of analysis undertaken by Radcliffe-Brown is described as concerned with 'the
relation of culture to the social groups who are its carriers'. Thus
'social structure' is merely an aspect of the all-inclusive phenomenon of 'culture'. It follows that there can be and must be a
single 'science of culture' which will embrace within a unitary
system of concepts and methods everything that makes up the
life of man in society. This is one point of view in the current
controversy. It is vigorously maintained by most American
and some Continental anthropologists, and would certainly
have been held by Malinowski.

The other point of view is more modest and corresponds to
Radcliffe-Brown's theoretical approach. It requires us to renounce
such grandiose aims and to accept the inevitability of a plurality
of frames of reference for the study of society. It may well be
that the concept of 'culture', as first formulated by Sir Edward
Tylor in *Primitive Culture* in 1871, identifies one such frame of
reference. It is certain that Radcliffe-Brown's concept of social
structure defines a specific frame of analysis which is not the same
as that intended by Dr Murdock and other American writers
when they speak of 'culture'.

This brings me back to Radcliffe-Brown's papers on ritual and
religion. It is of historical importance that Radcliffe-Brown's
major ethnographical work, *The Andaman Islanders*, which was
also his first full-scale demonstration of his theories, gave pride
of place to the analysis of ritual customs. This was not only
because he was then (1922) still greatly under the influence of
Durkheimian sociology. It was because Radcliffe-Brown discerned in the ritual customs of the Andamanese the basic mechanisms by which their social organization was maintained and
sanctioned. So he linked the contents of Andamanese beliefs with
the 'sentiments' required of them for the social order to be kept
going and with the 'valuations' made by them of their environmental resources for ensuring the survival of their society. In the
centre of the analysis stands the principle of the part they play in
maintaining the social organization as a whole; and this leads
to the view later more clearly expressed in the Myers Lecture
(1945) that religions vary 'in correspondence with the manner
in which the society is constituted'. It underlies the emphasis
on the relation between totemic institutions and segmentary

cleavages in the social structure which he develops in the essay on totemism; and it is the starting point of the concepts of 'ritual status' and 'ritual value' on which his Frazer lecture on Taboo turns.

Thus it is not what beliefs express in the symbolic, or intellectual, or emotional sense that concerns Radcliffe-Brown but what they do in defining the relations between persons and society or between society and nature. We are not concerned to establish whether a certain belief is magical (because it is based on fallacious association of ideas, assumes occult powers of acting on things and persons, and alleviates personal anxieties) or religious (because it involves the notion of superior beings, includes concepts of prayer and propitiation and is accompanied by moral rules and sanctions) as Frazer and Malinowski, and most other anthropologists, are. It is not for example the ostensible purpose of a totemic rite (e.g. the increase of the food supply), which we know to be futile, that is the main issue but the contribution made by the rite to the 'maintenance of that order of the universe of which man and nature are interdependent parts', which means, in the first place, the maintenance of the 'network of social relations binding individuals together in an ordered life'.

From this arise two rules of method in dealing with ritual usages in a single society, and by extension, for comparing ritual institutions in different societies. Instead of lumping together all the customs and beliefs found in a given 'culture' to which it seems fitting to give the label of magical, and similarly with other so-called categories of ritual beliefs and practices, the procedure is to compare all the different contexts and occasions in which a defined symbol appears or in which diverse symbolic ideas and actions are associated together. By this method of analysis we find that the occurrence of a birth and the occurrence of a death belong to the same class of ritual events. The Andamanese food and name taboos associated with these occurrences might, in Frazerian terms, just as satisfactorily be interpreted as the *cause* of the anxieties associated with such occurrences rather than as the ostensible defence against the anxieties. The method proposed avoids this. Birth and death are significant occasions of changes in the network of social relations. Those involved therefore undergo changes of ritual status and this is marked by the taboos they and other have to observe. As to the content of the taboos, this generally refers to objects like food in connection with which 'there continuously occur those inter-relations of interests which bind the individual men, women and children into a society'[1] and which are, consequently, endowed with ritual value. This is the connecting link between the social structure and the natural order within which every society has to live.

What I want to stress is the frame of analysis utilized. It is the same coherent and consistent system of concepts and method of relating and interpreting the data as is used in Radcliffe-Brown's studies of kinship, legal and political institutions. It is rigorously sociological, in the Durkheimian sense. Is taboo a form of negative magic due to the false logic and misdirected association of ideas of Frazer's ideal – and spurious – savage? Is it an illusory defence against the fears aroused by failure to cope in a practical way with the hazards of life, a form of psychological Dutch courage as Malinowski supposed? Is magic the result of the savage's pre-logical mentality, as Levy Bruhl maintained? Are taboos collective obsessions and paranoias as psychoanalytic writers argue? Are the Andamanese and Australian rites directed towards heavenly beings evidences of a degeneration from a primordial revelation of a Single Supreme Deity as Father Schmidt insisted? We need only make such a list of the more usual questions pursued by those who study primitive religion and magic to see how different Radcliffe-Brown's is; and, be it added, how free, by contrast, of unwarranted assumptions about how the mind, conscious and unconscious, of a hypothetical savage works or how his emotions are affected or how beliefs originated in the remote and undocumented past. But what this conceptual economy means is not that all other questions about ritual are irrelevant. It means that they should be formulated and investigated within other frames of analysis, with the corollary that to mix up frames of analysis is simply to cause confusion. This is now becoming well understood and we find psychologists, psychiatrists, and other specialist students undertaking field researches on the topics conventionally studied by anthropologists but using their specialist methods and concepts.[2] What is still not better understood than when Radcliffe-Brown gave his Frazer lecture in 1939, is that these specialist

[1] The quotations used in this outline of Radcliffe-Brown's theory of ritual come from the Frazer Lecture and the Myers Lecture.

[2] See, for example, the collection of papers made by C. Kluckhohn and H. A. Murray under the title of *Personality in Nature, Society and Culture* (1st edn., 1948, 2nd edn., 1953).

enquiries from whatever angle they are undertaken cannot provide the answers to the kind of questions raised by the frame of analysis used by Radcliffe-Brown and developed by some of his pupils and colleagues. It may well be that there is some sort of connection between a high incidence of food taboos in a particular society and child-rearing customs resulting in severe 'oral frustration'. We are still left with such problems as why the persons upon whom food taboos become incumbent differ from one social situation to another in the same society; and we still have no answer to the comparative problem of food taboos occurring in societies with widely diverse child-rearing customs. This is a simple, perhaps trivial illustration. It serves to make the point, though, that we must accept the inevitability of a plurality of frames of analysis and conceptual systems, at any rate for a considerable time to come, in the study of human society.

But methodology apart, it is also important to ask if Radcliffe-Brown's theories are substantive contributions to the formulation of generalizations about, and to the deeper understanding of, primitive ritual institutions. The answer is an unqualified affirmative. In the study of the classical subject of totemism a condition of chaos had been reached by the middle twenties. This was well shown in Goldenweiser's survey of the literature of the subject which led him to the conclusion that there is no such thing as totemism (Goldenweiser 1933). Radcliffe-Brown introduced a new way of handling the problem; and its first fruits were soon seen in such books as Warner's (1937), where Australian totemism is, for the first time, clearly exhibited in its relations to the social structure.

Even more impressive is Srinivas (1952). Here Radcliffe-Brown's methods and theories are ably applied to the religious system of a complex Oriental society. Professor Srinivas shows that there is as close an interlocking of religious institutions, which Frazer would certainly have placed at a higher 'stage' of evolution than Australian totemism, with the family and kinship system, the laws of inheritance and succession, and the principles determining the status of persons, among the Coorgs as among the Australians and Andamanese. He shows how the study of the 'higher' religions can be set free of the crippling trammels of theology, metaphysics and philology and brought within reach of social science.

In a field where such a diversity of views is found as in the study of ritual, judgements of what are significant methods and theories will differ. My own view is that Radcliffe-Brown's mark out the indispensable first steps in the scientific study of ritual customs and institutions in all societies. It is only *after* we have made the sort of analysis of ritual customs that is demonstrated in *The Andaman Islanders* that the way is opened for the psychological study of a ritual system.

At the beginning of this article I referred to the so-called British school of 'structuralist' anthropology and its debt to Radcliffe-Brown's teaching. The tendency thus labelled is most effectively represented in Professor E. E. Evans-Pritchard's distinguished book *The Nuer*, first published in 1940, which I have already spoken about. What is significant about this book is not only what it accomplishes but what it leaves out. It is an example of applying strict rules of relevance in the selection and presentation of field data in accordance with a clearly defined frame of analysis. It is ethnography based on a theoretical discipline, the discipline of structural analysis as first given shape by Radcliffe-Brown. An American student at the University of Chicago found the right word when he contrasted this kind of ethnography with what he called the 'grab-all' ethnography that is the conventional practice. This, in a nutshell, is our chief debt to Radcliffe-Brown. By following principles well proven in the natural sciences he has created the framework of a unified and rigorous discipline within the many-sided field of interests covered by the anthropological sciences. Like all scientific disciplines, it is limited to its proper range of data and works within the boundaries of its specific conceptual system. We can already see that the development of a science of social structure is bringing about rapid growth in such closely related disciplines as comparative jurisprudence and comparative politics. It will also make possible the much more rapid growth of the twin science of culture, whose main outlines have changed little since Tylor first set them down.

LIST OF WORKS CITED

Abbreviations used

A.A.   *American Anthropologist*
A.A.A. Mem. *Memoirs of the American Anthropological Association*
G.J.   *Geographical Journal*
I.L.R.   *Iowa Law Review*
J.A.F.   *Journal of American Folklore*
J.R.A.I.   *Journal of the Royal Anthropological Institute*
S.A.J.S.   *South African Journal of Science*
S.W.J.A.   *Southwestern Journal of Anthropology*
*cited by permission of the author

ASHTON, E. H., 1946. 'The Social Structure of the Southern Sotho Ward', *Communications from the School of African Studies*, N.S.15, Cape Town.

BARNES, J. A., 1960. 'African Models in the New Guinea Highland', *Man*, **62**, 2.
—— 1967. 'Agnation among the Enga: A Review Article', *Oceania*, **38**, 1, pp. 33–43.

BARTLETT, F. C., 1932. *Remembering: a study in experimental and social psychology*, pp. 199 ff., Cambridge.

BASCOM, W. R., 1944. 'The Sociological Role of the Yoruba Cult Group', *A.A.A. Mem.*, **63**.

BATESON, G., 1936. *Naven*, Cambridge.

BENEDICT, RUTH, 1935. *Patterns of Culture*, London.

BOHANNAN, LAURA, 1949. 'Dahomean Marriage: a revaluation', *Africa*, **19**, 4.
—— 1951. *A Comparative Study of Social Differentiation in Primitive Society*, (D.Phil. thesis, University of Oxford).*

BOHANNAN, P. J., 1951. *Political and Economic Aspects of Land Tenure and Settlement Patterns among the Tiv of Central Nigeria* (D.Phil. thesis, University of Oxford).*

BROWN, PAULA, 1962. 'Non-agnates among the Patrilineal Chimbu', *Jnl. Polynesian Society*, **71**, pp. 57–69.

BUCHLER, IRA R. and SELBY, HENRY A., 1968. *Kinship and Social Organization: An Introduction to Theory and Method*, New York.

BUSIA, K. A., 1951. *The Position of the Chief in the Modern Political System of Ashanti*, Oxford.

CAMBRIDGE ANCIENT HISTORY, 1924, vol. v., Cambridge.

CARSTAIRS, G. MORRIS, 1957. *The Twice-Born*, London.

COLLINGWOOD, R. G., 1944. *Autobiography*, Pelican Books, Harmondsworth.

COLSON, E., 1954. 'Ancestral Spirits and Social Structure among the Plateau Tonga', *Int. Arch. of Ethnogr.* **47**, pt. I, p. 21.

CONINGTON, J., 1872. *Virgil: a commentary*, London.

DANQUAH, J. B., 1928. *Akan Laws and Customs and the Akim Abuakwa Constitution*, London.

—— 1928b. *Cases in Akan Law*, London.

DAVENPORT, W., 1959. 'Non-unilinear Descent and Descent Groups', *A.A.*, **61**, pp. 557–72.

DE GROOT, J. J. M., 1910. *The Religion of the Chinese*, New York.

DEWEY, J., 1916. *Democracy and Education*, New York.

DOUGLAS, R. K., 1911. *Confucianism and Taoism*, London.

DRIBERG, J. H., 1932. *At Home with the Savage*, pp. 232 ff., London.

DRY, P. D. L., 1950. The Social Structure of a Hausa Village (B.Sc. thesis, University of Oxford).*

DUNNING, R. W., 1959. *Social and Economic Organization among the Northern Ojibway*, Toronto.

DURKHEIM, E., 1925. *L'Éducation morale*, Paris.

EGGAN, F., 1937. 'Cheyenne and Arapaho Kinship Systems', in *Social Organization of North American Tribes*, Chicago.

—— 1950. *The Social Organization of the Western Pueblos*, Chicago.

—— 1954. 'Social Anthropology and the Method of Controlled Comparison', *A.A.*, **56**, pp. 743–63.

EVANS-PRITCHARD, E. E., 1933–5. 'The Nuer: tribe and clan', *Sudan Notes and records*, **16**, Pt. 1, **17**, Pt. 2, **18**, Pt. 1.

—— 1937. *Witchcraft, Oracles and Magic among the Azande*, Oxford.

—— 1940a. *The Nuer*, Oxford.

—— 1940b. 'The Political System of the Nuer', in *African Political Systems* (M. Fortes and E. E. Evans-Pritchard, eds.), Oxford.

—— 1948. *The Divine Kingship of the Shilluk of the Nilotic Sudan*, Cambridge.

—— 1951a. *Kinship and Marriage among the Nuer*, Oxford.

—— 1951b. *Social Anthropology*, Oxford.

FIELD, M. J., 1940. *Social Organization of the Gã People*, Accra.

FIRTH, R., 1929. *Primitive Economics of the New Zealand Maori*, London.

—— 1936.*We, The Tikopia*, London.

—— 1946. *Malay Fishermen*, London.

—— 1951. 'Contemporary British Social Anthropology', *A.A.*, **53**, pp. 465–89.

—— 1954. 'Social Organization and Social Change', *J.R.A.I.*, **84**, pp. 1–20.

—— 1955. *The Fate of the Soul*, Cambridge.

—— 1957. 'A Note on Descent Groups in Polynesia', *Man*, **57**, 2.

—— 1959. *Social Change in Tikopia: A Re-Study of a Polynesian Community after a Generation*, London.

—— 1963. 'Bilateral Descent Groups: An Operational Viewpoint' in *Studies in Kinship and Marriage* (I. Schapera, ed.), Occasional Papers of the Royal Anthropological Institute, London.

FLUGEL, J. C., 1921. *The Psychoanalytic Study of the Family*, London.

FORDE, C. D., 1935. 'On the Concept of Function', *A.A.*, **37**, 3.

—— 1938. 'Fission and Accretion in the Patrilineal Clans of a Semi-Bantu Community in Southern Nigeria', *J.R.A.I.*, **68**, pp. 311–38.

—— 1939. 'Kinship in Umor: Double Unilateral Organization in a Semi-Bantu Society', *A.A.*, **41**.

—— 1941. *Marriage and the Family among the Yakö in South-Eastern Nigeria*,

London School of Economics Monographs on Social Anthropology, no. 5, London.

—— 1947. 'The Anthropological Approach in the Social Sciences', Presidential address, Sec. H., Brit. Assoc., *The Advancement of Science*, **15**, p. 213.

—— 1950a. 'Double Descent among the Yakö, in *African Systems of Kinship and Marriage* (A. R. Radcliffe-Brown and C. D. Forde, eds.), Oxford.

—— 1950b. 'Ward Organization among the Yakö', *Africa*, **20**, 4.

—— 1951. 'The Yoruba-Speaking Peoples of South-Western Nigeria', *Ethnographic Survey of Africa*, Pt. IV, London.

—— and Jones, G. I., 1950. 'The Ibo and Ibibio-Speaking Peoples of South-Eastern Nigeria', *Ethnographic Survey of Africa*, *Western Africa*, Pt. III, London.

FORTES, M., and S. L., 1936. 'Food in the Domestic Economy of the Tallensi', *Africa*, **9**, no. 2.

FORTES, M., 1936a. 'Ritual Festivals and Social Cohesion in the Hinterland of the Gold Coast', *A.A.*, **38**, no. 4. (Reprinted in this volume as ch. 6.)

—— 1936b. 'Kinship, Incest and Exogamy in the Northern Territories of the Gold Coast', in *Custom is King*, Essays presented to R. R. Marett (L. H. D. Buxton, ed.), London.

—— 1940. 'The Political System of the Tallensi of the Northern Territories of the Gold Coast' in *African Political Systems* (M. Fortes and E. E. Evans-Pritchard, eds.), Oxford.

—— 1945. *The Dynamics of Clanship among the Tallensi*, Oxford.

—— 1949a. *The Web of Kinship among the Tallensi*, London.

—— 1949b. 'Time and Social Structure: An Ashanti Case Study' in *Social Structure: Studies presented to A. R. Radcliffe-Brown* (M. Fortes, ed.), Oxford. (Reprinted in this volume as ch. 1.)

—— 1950. 'Kinship and Marriage among the Ashanti', in *African Systems of Kinship and Marriage* (A. R. Radcliffe-Brown and C. D. Forde, eds.), Oxford.

—— 1953a. 'Analysis and Description in Social Anthropology', Presidential address, Sec. H., Brit. Assoc., *The Advancement of Science*, **38**. (Reprinted in this volume as ch. 5.)

—— 1953b. 'The Structure of Unilineal Descent Groups', *A.A.*, **55**, no. 1. (Reprinted in this volume as ch. 3.)

—— 1954. 'A demographic field study in Ashanti' in *Culture and Human Fertility* (Frank Lorimer, ed.), Unesco, Paris.

—— 1955. 'Radcliffe-Brown's Contributions to the Study of Social Organization', *Br. Jnl. of Sociology*, **6**, Pt. 1. (Reprinted in this volume as ch. 9.)

—— 1957. 'Malinowski and the Study of Kinship', in *Man and Culture: an Evaluation of the Works of Bronislaw Malinowski* (R. Firth, ed.), London.

—— 1958. Introduction to 'The Developmental Cycle in Domestic Groups', *Cambridge Papers in Social Anthropology*, no. 1 (J. Goody, ed.).

—— 1959. *Oedipus and Job in West African Religion*, Cambridge.

—— 1964. 'Some Reflections on Ancestor Worship in Africa' in *African Systems of Thought*, preface by M. Fortes and G. Dieterlen, London.

—— 1969 forthcoming). *Kinship and the Social Order: The Legacy of Lewis Henry Morgan*, Chicago.

FORTES, M., and KYEI, T. E., 1945. Unpublished Field Data of the Ashanti Social Survey, referred to in A. Phillips, L. Mair and L. Harries, 1952, *Survey of African Marriage and Family Life*, London.

—— STEEL, R. W., and ADY, P., 1948. 'Ashanti Survey, 1945–6: An Experiment in Social Research', *G.J.*, **110**, 4–6, 149–79.

FORTUNE, R. F., 1932. *Sorcerers of Dobu*, London.

FOWLER, W. WARDE, 1911. *The Religious Experience of the Roman People*, London.

FREEDMAN, M., 1958. *Lineage Organization in Southeastern China*, London School of Economics Monographs on Social Anthropology, no. 18, London.

—— 1966. *Chinese Lineage and Society: Fukien and Kwangtung*, London School of Economics Monographs on Social Anthropology, no. 33, London.

FREEMAN, J. D., 1958. 'The Family System of the Iban of Borneo', *Cambridge Papers in Social Anthropology*, no. 1 (J. Goody, ed.).

—— 1961. 'On the Concept of the Kindred', *J.R.A.I.*, **91**, pp. 192–220.

FRIED, MORTON, H., 1957. 'The Classification of Corporate Unilineal Descent Groups', *J.R.A.I.*, **87**, pp. 1–29.

FUSTEL DE COULANGES, 1864. *La Cité Antique*, Paris.

GENNEP, A. VAN, 1909. *Les Rites de Passage*, Paris.

GLUCKMAN, M., 1950. 'Kinship and Marriage among the Lozi of Northern Rhodesia and the Zulu of Natal', in *African Systems of Kinship and Marriage* (A. R. Radcliffe-Brown and C. D. Forde, eds.), Oxford.

—— 1951. 'The Lozi of Barotseland in North Western Rhodesia' in *Seven Tribes of British Central Africa* (E. Colson and M. Gluckman, eds.), London.

GOLDENWEISER, A. A., 1933. *History, Psychology and Culture*, London.

GOODENOUGH, WARD A., 1951. *Property, Kin, and Community on Truk*, Yale University Publications in Anthropology, 46.

—— 1955. 'A Problem in Malayan and Polynesian Social Organization', *A.A.*, **57**, pp. 71–83.

GOODY, JACK (ed.), 1958a. 'The Developmental Cycle in Domestic Groups', *Cambridge Papers in Social Anthropology*, no. 1, Cambridge.

—— 1958b. 'The Fission of Domestic Groups among the LoDagaba', *Cambridge Papers in Social Anthropology*, no. 1 (J. Goody, ed.), Cambridge.

—— 1959. 'The Mother's Brother and Sister's Son in West Africa', *J.R.A.I.*, **89**.

—— 1961. 'The Classification of Double Descent Systems', *Current Anthropology*, **2**, pp. 3–25.

—— 1962. *Death, Property and the Ancestors*, Stanford.

GOUGH, E. KATHLEEN, 1950. *Kinship among the Nayar of the Malabar Coast of India* (Ph.D. thesis, University of Cambridge).★

—— 1952. 'Changing Kinship Usages in the Setting of Political and Economic Change among the Nayars of Malabar', *J.R.A.I.*, **82**.

—— 1958. 'Cults of the Dead among the Nayar', *J.A.F.*, **71**, 281.

—— 1961. 'Nayar: Central Kerala' in *Matrilineal Kinship* (D. Schneider and Kathleen Gough, eds.), Berkeley.

GRANET, MARCEL, 1951. *La Religion des Chinois* (2nd edn.), Paris.

GRAY, R. F. and GULLIVER, P. H. (eds.), 1964. *The Family Estate in Africa: Studies of the Role of Property in Family Structure and Lineage Continuity*, London.

GRUBE, WILHELM, 1910. *Religion und Kultus der Chinesen*, Berlin.

HADDON, A. C., 1905. 'Illustrations of Folk Lore', Presidential Address, See. H., Brit. Assoc. Adv. Sci., *Annual Report*.

HELSER, A. D., 1934. *Education of Primitive People*, New York.

HERSKOVITS, M. J., 1938. *Dahomey*, New York.

—— 1949. *Man and His Works*, New York.

HOERNLÉ, A. W., 1931. 'An Outline of the Native Conception of Education in Africa', *Africa*, **4.**

HOMANS, G. and SCHNEIDER, D., 1955. *Marriage, Authority and Final Causes*, Illinois.

HUNTER, MONICA, 1936. *Reaction to Conquest*, London.

HSU, F. L. K., 1949. *Under the Ancestors' Shadow*, London.

IRVINE, A. L. (ed.), 1924. *The Fourth Book of Virgil's Aeneid*, Oxford.

ISAACS, S., 1930. *Intellectual Growth in Young Children*, London.

—— 1951. *Social Development in Young Children*, London.

JENNINGS, W. IVOR, 1939, reprinted 1948. *Parliament*, Cambridge.

JUNOD, H., 1927. *The Life of a South African Tribe*, 2 vols., London.

KAUFMANN, F., 1944. *Methodology of the Social Sciences*, London.

KLUCKHOHN, C., and MURRAY, H. A., eds., 1948, 2nd edn. 1953. *Personality in Nature, Society and Culture*, New York.

KOFFKA, K., 1928, 2nd edn. *Growth of the Mind*, London.

KROEBER, A. L., 1938. 'Basic and Secondary Patterns of Social Structure', *J.R.A.I.*, **68.**

—— 1952. *The Nature of Culture*, chs. 23 and 24, Chicago.

KUPER, HILDA, 1947. *An African Aristocracy*, London.

—— 1950. 'Kinship among the Swazi', in *African Systems of Kinship and Marriage* (A. R. Radcliffe-Brown and C. D. Forde, eds.), Oxford.

LABOURET, H., 1958. *Les Tribus du Rameau Lobi*, Dakar.

LA FONTAINE, JEAN, 1960. 'Homicide and Suicide among the Gisu', in *African Homicide and Suicide* (P. Bohannan, ed.), Princeton.

LANGNESS, LEWIS L., 1964. 'Some Problems in the Conceptualisation of Highlands Social Structure' in *New Guinea, A.A.*, Special Publication, **66,** 4, Pt. 2, (James B. Watson, ed.), pp. 162–82.

LAYARD, J., 1942. *Stone Men of Malekula*, London.

LEACH, E. R., 1951. 'The Structural Implications of Cross-Cousin Marriage', *J.R.A.I.*, **81.**

—— 1954. *Political Systems of Highland Burma*, London.

—— 1957. 'Aspects of Bridewealth and Marriage Stability among the Kachin and Lakher', *Man*, **59.** Reprinted in *Rethinking Anthropology*, London School of Economics Monographs on Social Anthropology, no. 22, London.

—— 1958. 'Magical Hair', *J.R.A.I.*, **88,** 147.

—— 1960. 'The Sinhalese of the Dry Zone of Northern Ceylon' in *Social Structure in South East Asia*, Viking Fund Publications in Anthropology, no. 29, pp. 116–26.

—— 1961a. *Pul Eliya: A Village in Ceylon*, Cambridge.

—— 1961b. *Rethinking Anthropology*, London School of Economics Monographs on Social Anthropology, no. 22, London.

LEACH, E. R., 1962. 'On Certain Unconsidered Aspects of Double Descent', *Man*, **62**, 214.

LEVI-STRAUSS, C., 1949. *Les Structures Élémentaires de la Parenté*, Paris.

LEWIS, CHARLTON T., and SHORT, CHARLES, 1945. *Latin-English Dictionary*, Oxford.

LIN, YUEH-HWA, 1948. *The Golden Wing: a Sociological Study of Chinese Familism*, London.

LOWIE, R. H., 1921. *Primitive Society*, London.

MACIVER, R. M., 1937. *Society*, London.

MAINE, Sir HENRY S., 1861. *Ancient Law*, London.

MALINOWSKI, B., 1927. *The Father in Primitive Psychology*, London.

—— 1929. *The Sexual Life of Savages*, London.

—— 1932a. *Argonauts of the Western Pacific*, London.

—— 1932b. *Crime and Custom in Savage Society*, New York.

—— 1936. 'Native Education and Culture Contact', *International Review of Missions*.

MAYER, P., 1949. 'The Lineage Principle in Gusii Society', *International African Institute, Memorandum XXIV*.

—— 1950. *Gusii Bridewealth Law and Custom*, Rhodes-Livingstone Papers, no. 18, Livingstone.

MEAD, MARGARET, 1928. *Coming of Age in Samoa*, New York.

—— 1931. *Growing up in New Guinea*, London.

—— 1934. *Kinship in the Admiralty Islands*, American Museum of Natural History.

MEEK, C. K., 1937. *Law and Authority in a Nigerian Tribe*, London.

MEGGITT, M. J., 1965. *The Lineage System of the Mae-Enga of New Guinea*, Edinburgh.

MIDDLETON, JOHN, 1960. *Lugbara Religion*, London.

MITCHELL, J. CLYDE, 1950. *Social Organization of the Yao of Southern Nyasaland* (D.Phil. thesis, University of Oxford).*

—— 1951. 'The Yao of Southern Nyasaland', in *Seven Tribes of British Central Africa* (E. Colson and M. Gluckman, eds.), London.

—— 1956. *The Yao Village*, Manchester.

MONTEIL, C., 1924. *Les Bambara du Ségou et du Kaarta*, Paris.

MUMFORD, W. B., 1930. 'Malangali Experiments', *Africa*, **3**.

MURDOCK, G. P., 1949. *Social Structure*, New York.

—— 1951. 'British Social Anthropology', *A.A.*, **53**, pp. 465–89.

—— 1960. 'Cognatic Forms of Social Organization', in *Social Structure in Southeast Asia* (G. P. Murdock, ed.), Viking Fund Publications in Anthropology, no. 29, pp. 1–14, Chicago.

MURPHY, G., 1931. *Experimental Social Psychology*, New York.

NADEL, S. F., 1942. *A Black Byzantium*, London.

—— 1950. 'Dual Descent in the Nuba Hills', in *African Systems of Kinship and Marriage* (A. R. Radcliffe-Brown and C. D. Forde, eds.), Oxford.

—— 1951. *Foundations of Social Anthropology*, London.

OBERG, K., 1940. 'The Kingdom of Ankole in Uganda', in *African Political Systems* (M. Fortes and E. E. Evans-Pritchard, eds.), Oxford.

PARRY, N. E., 1932. *The Lakhers*, London.

PAULY, A. F. VON, 1950. *Real-Encyclopaedie der Klassischen Altertumswissenschaft*, Stuttgart.

PEHRSON, R. N., 1957. *The Bilateral Network of Social Relations in Könkama Lapp District*, Indiana University Publications, Slavic and East European Series, 5.

PETERS, E.L., 1951. *The Sociology of the Beduin of Cyrenaica* (D.Phil. thesis, University of Oxford).*

PHILLIPS, A., MAIR, L., and HARRIES, L., 1952. *Survey of African Marriage and Family Life*, London.

PIAGET, JEAN, 1926. *The Language and Thought of the Child*, London.

RADCLIFFE-BROWN, A. R., 1913. 'Three Tribes of Western Australia', *J.R.A.I.*, **43.**

—— 1922. *The Andaman Islanders*, Cambridge.

—— 1923. 'Methods of Ethnology and Social Anthropology', *S.A.J.S.*, **20.**

—— 1924. 'The Mother's Brother in South Africa', *S.A.J.S.*, **21,** pp. 542–55.

—— 1929. 'The Sociological Theory of Totemism', *Proceedings of the Fourth Pacific Science Congress*, Java.†

—— 1930. 'The Social Organization of Australian Tribes', *Oceania*, no. 1, Melbourne.

—— 1931. 'The Present Position of Anthropological Studies', Presidential Address, Sec. H, Brit. Assoc. Adv. Sci., *Annual Report*.

—— 1933a. 'Primitive Law', *Encyclopaedia of the Social Sciences*, **9,** pp. 202–6, New York.†

—— 1933b. 'Social Sanctions', *Encyclopaedia of the Social Sciences*, **13,** pp. 531–4, New York.†

—— 1935a. 'Patrilineal and Matrilineal Succession', *I.L.R.*, **20,** 2.†

——1935b. 'On the Concept of Function in Social Science', *A.A.*, **37,** p. 3.†

—— 1939. 'Taboo', *The Frazer Lecture*, Cambridge.†

—— 1940a. 'On Social Structure', Presidential Address to the Royal Anthropological Institute, *J.R.A.I.*, **70.**†

—— 1940b. 'On Joking Relationships', *Africa*, **13,** 3, pp. 195–210.†

—— 1940c. Preface to *African Political Systems* (M. Fortes and E. E. Evans-Pritchard, eds.), Oxford.

—— 1941. 'The Study of Kinship Systems', Presidential Address to the Royal Anthropological Institute, *J.R.A.I.*, **71.**†

—— 1945. 'Religion and Society', *The Henry Myers Lecture*, *J.R.A.I.*, **75.**†

—— 1949. 'A Further Note on Joking Relationships', *Africa*, **19,** pp. 133–40.†

—— 1950. Introduction to *African Systems of Kinship and Marriage* (A. R. Radcliffe-Brown and C. D. Forde, eds.), Oxford.

—— 1951. 'The Comparative Method in Social Anthropology', *J.R.A.I.*, **81.**

—— 1952. *Structure and Function in Primitive Society*, a collection of articles, including those above, marked †, London.

RATTRAY, R. S., 1923. *Ashanti*, Oxford.

—— 1927. *Religion and Art in Ashanti*, Oxford.

—— 1929. *Ashanti Law and Constitution*, Oxford.

—— 1932. *Tribes of the Ashanti Hinterland*, Oxford.

RICHARDS, A. I., 1934. 'Mother Right among the Central Bantu', in *Essays presented to C. G. Seligman* (E. E. Evans-Pritchard, Raymond Firth, B. Malinowski and I. Schapera, eds.), London.

―― 1939. *Land, Labour and Diet in Northern Rhodesia*, London.

――1940a. *Bemba Marriage and Present Economic Conditions*, Rhodes-Livingstone Institute Papers, no. 4, Livingstone.

―― 1940b. 'The Political System of the Bemba Tribe, north-eastern Rhodesia', in *African Political Systems* (M. Fortes and E. E. Evans-Pritchard, eds.), Oxford.

―― 1950. 'Some Types of Family Structure among the Central Bantu', in *African Systems of Kinship and Marriage* (A. R. Radcliffe-Brown and C. D. Forde, eds.), Oxford.

RIVERS, W. H. R., 1906. *The Todas*, London.

―― 1911. 'The Ethnological Analysis of Culture', Presidential Address, Sec. H., Brit. Assoc. Adv. Sci., *Annual Report*.

―― 1941a. *Kinship and Social Organization*, London.

――1914b. *The History of Melanesian Society*, 2 vols., Cambridge.

―― 1922. 'The Unity of Anthropology', *J.R.A.I.*, **52**, pp. 12 ff.

ROBINSON, MARGUERITE S., 1962. 'Complementary Filiation and Marriage in the Trobriand Islands' in *Marriage in Tribal Societies* (M. Fortes, ed.), *Cambridge papers in Social Anthropology*, no. 3, pp. 121–55, Cambridge.

ROSCHER, W. H., 1902–9. *Lexikon der Griechischen und Römischen Mythologie*, Leipzig.

SCHAPERA, I., 1935. 'Social Structure of the Tswana Ward', *Bantu Studies*, **9**, no. 3.

―― 1938 (2nd edn. 1955). *Handbook of Tswana Law and Custom*, London.

―― 1940. *Married Life in an African Tribe*, London.

―― 1943. *Native Land Tenure in the Bechuanaland Protectorate*, Lovedale.

―― 1950. 'Kinship and Marriage among the Tswana' in *African Systems of Kinship and Marriage* (A. R. Radcliffe-Brown and C. D. Forde, eds.), Oxford.

SCHEFFLER, H., 1966. 'Ancestor Worship in Anthropology: or Observations on Descent and Descent Groups', *Current Anthropology*, **7**, pp. 541–51.

SCHNEIDER, DAVID M., 1965. 'Some Muddles in the Models or How the System Really Works', in *The Relevance of Models for Social Anthropology* (Michael Banton, ed.), pp. 25–85, London.

―― 1967. 'Descent and Filiation as Cultural Constructs', *S.W.J.A.*, **23**, 1, pp. 65–73.

SELIGMAN, C. G., 1915. 'On the Physical Characters of the Ancient Egyptians', Presidential Address, Sec. H., Brit. Assoc. Adv. Sci., *Annual Report*.

―― 1932. 'Anthropological Perspective and Psychological Theory', *J.R.A.I.*, **62**, 193.

―― and SELIGMAN, B. Z., 1932. *Pagan Tribes of the Nilotic Sudan*, London.

SIDER, KAREN B., 1967. 'Kinship and Culture: Affinity and the Role of the Father in the Trobriands', *S.W.J.A.*, **23**, 1, pp. 90–109.

SIMEY, T. S., 1946. *Welfare and Planning in the West Indies*, Oxford.

SMITH, E. W., 1934. 'Africa: what do we know of it?', *J.R.A.I.*, **65**.

―― 1935. 'Indigenous Education in Africa', in *Essays presented to C. G.*

Seligman (E. E. Evans-Pritchard, Raymond Firth, B. Malinowski and I. Schapera, eds.), London.

—— 1952. 'African Symbolism', *J.R.A.I.*, **82**, pp. 13–37.

SMITH, M. G., 1956. 'Segmentary Lineage Systems', *J.R.A.I.*, **86**, 2, pp. 39–80.

SMITH, RAYMOND T., 1956. *The Negro Family in British Guiana*, London.

SMITH, W. ROBERTSON, 1889. *The Religion of the Semites*, 3rd edn. 1938, Cambridge.

SOUTHALL, AIDAN, 1956. *Alur Society: A Study in Processes and Types of Domination*, Cambridge.

SPEARMAN, C. E., 1923. *Nature of Intelligence*, London.

SPOEHR, A., 1947. 'Changing Kinship Systems', Anthropological Series, Chicago Natural History Museum, **33**, 4.

—— 1950. 'Observations on the Study of Kinship', *A.A.*, **52**.

SRINIVAS, M. N., 1952. *Religion and Society among the Coorgs of South India*, Oxford.

STENNING, D. J., 1958. 'Household Viability among the Pastoral Fulani', *Cambridge Papers in Social Anthropology*, no. 1 (J. Goody, ed.), Cambridge.

SU, SING GING, 1922. *The Chinese Family System* (Ph.D. thesis, Columbia University, N.Y.).★

TYLOR, Sir E., 1871. *Primitive Culture*, London.

WALEY, ARTHUR, 1938. *The Analects of Confucius*, London.

WARNER, W. L., 1937. *A Black Civilization*, New York.

WEBER, MAX, 1947. *The Theory of Social and Economic Organization*, translated by A. R. Henderson and Talcott Parsons, London.

WESTRUP, C. W., 1939–44. *Introduction to Early Roman Law*, 5 vols., Cambridge.

WHITE, LESLIE, 1947. 'Evolutionary Stages, Progress and the Evaluation of Cultures', *S.W.J.A.*, **3**, no. 3, pp. 184–5.

WILSON, G. and MONICA, 1945. *The Analysis of Social Change*, Cambridge.

WILSON, MONICA, 1950. 'Nyakyusa Kinship', in *African Systems of Kinship and Marriage* (A. R. Radcliffe-Brown and C. D. Forde, eds.), Oxford.

—— 1951a. 'Nyakyusa Age Villages', *J.R.A.I.*, **81**.

—— 1951b. *Good Company: A Study of Nyakyusa Age Villages*, London.

YANG, M. C., 1945. *A Chinese Village*, London.

LONDON SCHOOL OF ECONOMICS
MONOGRAPHS ON SOCIAL ANTHROPOLOGY

Titles marked with an asterisk are now out of print. Those marked with a dagger have been reprinted in paperback editions and are only available in this form. A double dagger indicates availability in both hardcover and paperback editions.

†19. FREDRIK BARTH
Political Leadership among Swat Pathans, 1959.

†20. L. H. PALMIER
Social Status and Power in Jawa, 1960.

†21. JUDITH DJAMOUR
Malay Kinship and Marriage in Singapore, 1959.

†22. E. R. LEACH
Rethinking Anthropology, 1961.

★23. S. M. SALIM
Marsh Dwellers of the Euphrates Delta, 1962.

†24. S. VAN DER SPRENKEL
Legal Institutions in Manchu China, 1962.

25. CHANDRA JAYAWARDENA
Conflict and Solidarity in a Guianese Plantation, 1963.

26. H. IAN HOGBIN
Kinship and Marriage in a New Guinea Village, 1963.

27. JOAN METGE
A New Maori Migration: Rural and Urban Relations in Northern New Zealand, 1964.

‡28. RAYMOND FIRTH
Essays on Social Organization and Values, 1964.

29. M. G. SWIFT
Malay Peasant Society in Jelebu, 1965.

†30. JEREMY BOISSEVAIN
Saints and Fireworks: Religion and Politics in Rural Malta, 1965.

31. JUDITH DJAMOUR
The Muslim Matrimonial Court in Singapore, 1966.

32. CHIE NAKANE
Kinship and Economic Organization in Rural Japan, 1967.

33. MAURICE FREEDMAN
Chinese Lineage and Society: Fukien and Kwangtung, 1966.

34. W. H. R. RIVERS
Kinship and Social Organization, reprinted with commentaries by David Schneider and Raymond Firth, 1968.

35. ROBIN FOX
The Keresan Bridge: A Problem in Pueblo Ethnology, 1967.

36. MARSHALL MURPHREE
Christianity and the Shona, 1969.

37. G. K. NUKUNYA
Kinship and Marriage among the Anlo Ewe, 1969.

38. LUCY MAIR
Anthropology and Social Change, 1969.

39. SANDRA WALLMAN
Take Out Hunger: Two Case Studies of Rural Development in Basutoland, 1969.

40. MEYER FORTES
Time and Social Structure and Other Essays, 1970.

41. J. D. FREEMAN
Report on the Iban, 1970.

42. W. E. WILLMOTT
The Political Structure of the Chinese Community in Cambodia, 1970.

43. I. SCHAPERA
Tribal Innovators: Tswana Chiefs and Social Change 1795–1940, 1970.